e-Learning in Medical Physics and Engineering

Series in Medical Physics and Biomedical Engineering

Series Editors:
Russell Ritenour, Slavik Tabakov and Kwan-Hoong Ng

RECENT BOOKS IN THE SERIES

e-Learning in Medical Physics and Engineering:
Building Educational Modules with Moodle
Vassilka Tabakova

Proton Therapy Physics, Second Edition
Harald Paganetti (Ed)

Mixed and Augmented Reality in Medicine
Terry M. Peters, Cristian A. Linte, Ziv Yaniv, Jacqueline Williams (Eds)

Graphics Processing Unit-Based High Performance
Computing in Radiation Therapy
Xun Jia, Steve B. Jiang (Eds)

Clinical Radiotherapy Physics with MATLAB:
A Problem-Solving Approach
Pavel Dvorak

Advanced and Emerging Technologies in Radiation
Oncology Physics
Siyong Kim, John W. Wong (Eds)

Advances in Particle Therapy: A Multidisciplinary Approach
Manjit Dosanjh, Jacques Bernier (Eds)

For more information about this series, please visit: https://
www.crcpress.com/Series-in-Medical-Physics-and-Biomedical-
Engineering/book-series/CHMEPHBIOENG

e-Learning in Medical Physics and Engineering

Building Educational Modules with Moodle

Vassilka Tabakova

CRC Press
Taylor & Francis Group
Boca Raton London New York

CRC Press is an imprint of the
Taylor & Francis Group, an **Informa** business

First edition published 2020
by CRC Press
6000 Broken Sound Parkway NW, Suite 300, Boca Raton, FL 33487-2742

and by CRC Press
2 Park Square, Milton Park, Abingdon, Oxon, OX14 4RN

First issued in paperback 2021

© 2020 Taylor & Francis Group, LLC
CRC Press is an imprint of Taylor & Francis Group, an Informa business

ISBN 13: 978-1-03-224369-6 (pbk)
ISBN 13: 978-1-138-34732-8 (hbk)

DOI: 10.1201/9780429437052

Publisher's Note
The publisher has gone to great lengths to ensure the quality of this reprint but points out that some imperfections in the original copies may be apparent.

Visit the Taylor & Francis Web site at
http://www.taylorandfrancis.com

and the CRC Press Web site at
http://www.crcpress.com

Typeset in Minion
by codeMantra

Contents

About the Series

THE SERIES IN MEDICAL *Physics and Biomedical Engineering* describes the applications of physical sciences, engineering, and mathematics in medicine and clinical research.
The series seeks (but is not restricted to) publications in the following topics:

- Artificial organs

- Assistive technology

- Bioinformatics

- Bioinstrumentation

- Biomaterials

- Biomechanics

- Biomedical engineering

- Clinical engineering

- Imaging

- Implants

- Medical computing and mathematics

- Medical/surgical devices

- Patient monitoring

- Physiological measurement

- Prosthetics

- Radiation protection, health physics, and dosimetry

- Regulatory issues

- Rehabilitation engineering

- Sports medicine

- Systems physiology

- Telemedicine

- Tissue engineering

- Treatment

The *Series in Medical Physics and Biomedical Engineering* is an international series that meets the need for up-to-date texts in this rapidly developing field. Books in the series range in level from introductory graduate textbooks and practical handbooks to more advanced expositions of current research.

The *Series in Medical Physics and Biomedical Engineering* is the official book series of the International Organization for Medical Physics.

THE INTERNATIONAL ORGANIZATION FOR MEDICAL PHYSICS

The International Organization for Medical Physics (IOMP) represents over 18,000 medical physicists worldwide and has a membership of 80 national and 6 regional organizations, together with a number of corporate members. Individual medical physicists of all national member organisations are also automatically members.

The mission of IOMP is to advance medical physics practice worldwide by disseminating scientific and technical information, fostering the educational and professional development of medical physics and promoting the highest quality medical physics services for patients.

A World Congress on Medical Physics and Biomedical Engineering is held every three years in cooperation with International Federation for Medical and Biological Engineering (IFMBE) and International Union for Physics and Engineering Sciences in Medicine (IUPESM). A regionally based international conference, the International Congress of Medical Physics (ICMP) is held between world congresses. IOMP also sponsors international conferences, workshops and courses.

The IOMP has several programmes to assist medical physicists in developing countries. The joint IOMP Library Programme supports 75 active libraries in 43 developing countries, and the Used Equipment Programme coordinates equipment donations. The Travel Assistance Programme provides a limited number of grants to enable physicists to attend the world congresses.

IOMP co-sponsors the *Journal of Applied Clinical Medical Physics*. The IOMP publishes, twice a year, an electronic bulletin, *Medical Physics World*. IOMP also publishes e-Zine, an electronic news letter about six times a year. IOMP has an agreement with Taylor & Francis for the publication of the *Medical Physics and Biomedical Engineering* series of textbooks. IOMP members receive a discount.

IOMP collaborates with international organizations, such as the World Health Organisations (WHO), the International Atomic Energy Agency (IAEA) and other international professional bodies such as the International Radiation Protection Association (IRPA) and the International Commission on Radiological Protection (ICRP), to promote the development of medical physics and the safe use of radiation and medical devices.

Guidance on education, training and professional development of medical physicists is issued by IOMP, which is collaborating with other professional organizations in development of a professional certification system for medical physicists that can be implemented on a global basis.

The IOMP website (www.iomp.org) contains information on all the activities of the IOMP, policy statements 1 and 2 and the 'IOMP: Review and Way Forward' which outlines all the activities of IOMP and plans for the future.

Acknowledgements

I MOST GRATEFULLY ACKNOWLEDGE the contribution of the international Consortia of the pioneering projects in Medical Physics and Engineering described in the book and especially their coordinator, Prof. Slavik Tabakov, for creating the environment of e-Learning in the profession, in which I could gain the experience necessary to write this book.

I am also very grateful to my family for their valuable support throughout the process of writing this book.

I am very grateful to the International Centre of Theoretical Physics (ICTP) and the University of Trieste and especially the management of their College on Medical Physics and the Master of Advanced Studies in Medical Physics programme, who made possible for me to conduct surveys and test parts of this guide at some of the sessions there.

I am also grateful to the organisers of the various international forums on Medical Physics and Engineering for opportunities to present my work related to e-Learning, and especially to Prof. Anchali Krisanachinda and Prof. Arun Chougule for their support when conducting the surveys at these forums.

I am also grateful to Prof. Magdalena Stoeva, editor of CRC Focus series, who encouraged me to disseminate my experience in e-Learning in the form of this book. My thanks also go to the editorial and production team at CRC Press/Taylor & Francis Group, especially Rebecca Davies, Kirsten Barr, Todd Perry and codeMantra's Saranya P N and the team, for their unfailing support and direction.

I am also grateful to my colleagues at King's College London, in particular to the departments of Medical Engineering and Physics, Dentistry Distance Learning and the Virtual Campus & Technology Enhanced Learning Unit.

As are all users, I am grateful to the creators of Moodle for the favourable conditions to establish the VLE platform described in the book.

During the years, I have had the pleasure and privilege of working with over 300 colleagues – medical physicists and engineers – from around the world, who contributed to the pioneering e-Learning projects briefly described in this book. To witness first hand their enthusiasm when developing stages of those projects has been in itself a great moral support and has convinced me of the value of writing this book.

Introduction

e-LEARNING IN MEDICAL PHYSICS AND ENGINEERING: *Building Educational Modules with Moodle* is your guide to using the most popular free and open-source Moodle Virtual Learning Environment (VLE) platform in Higher Education for application in any area with a considerable image-based educational component. Whether you aim is to build your own e-Learning course or a whole education Programme, the author believes that you will find this guide reliable, intuitive, and user-friendly. The working examples will make your task easy and enjoyable.

WHO IS THE BOOK FOR?

Primarily the book was designed for specialists in Medical Physics and Engineering as, nowadays, it is impossible to imagine contemporary medicine without the support of physics and technology. Some of the greatest inventions in physics such as X-rays, nuclear magnetic resonance, ultrasound, particle accelerators, and radioisotope tagging and detection techniques have brought revolutionary changes to medicine to create sophisticated diagnostic and treatment equipment and methods.

The need for qualified specialists to work with and apply these sophisticated types of equipment, methods, and technology is rapidly growing. Professional bodies predict the necessity of almost tripling the number of Medical Physicists by 2035, with the most significant growth being expected in lower- and middle-income countries. Similar pressures exist in Medical Engineering, which is a constantly growing profession, especially now with the rapid introduction of Hospital Information Systems and Artificial Intelligence. Efficient and fast growth of the number of these specialists will have to rely on broad implementation of e-Learning. A free and open-source Virtual Learning Platform, such as Moodle, will be very useful for providing this on a limited budget. However, because organising an e-Learning platform, no matter whether it is free and open-source or proprietary, is subject to the general principles that have been reflected in this book. The author believes that the present guide will be useful not just to Moodle users, but also to a much broader audience.

As the book evolved, it appeared that it will appeal to every educator interested in the quick application of e-Learning in their work not only in the field of Medical Physics and Engineering, but also in medical or medical-related specialties and others. The organisation of e-Learning platforms and creation of e-Learning materials in these professions need to take into consideration the fact that all these rely heavily on images. Thus, the book will be of interest also to educators in areas of the Humanities and other disciplines where there is a strong imaging component.

Most importantly, it should be noted that no specialist IT knowledge and no acquaintance with programming are required to use this guide. This book provides a condensed step-by-step guide for the professional educator, who has to quickly build an e-Learning course with Moodle starting from scratch. For the user to have an overview of this process, illustrations in the book are given as screen-shots of the whole screen, rather than of single details and explanations follow in the text. As the illustrations are generated from a free version of Moodle, it is possible for readers to make their own free trials following the steps in Chapter 5, especially if there is need for viewing more details. Additionally there are specific detailed books for Moodle addressing IT specialists in the field – developers, system administrators, etc. as well as general users, which can be found online or on the Internet and traditional booksellers.

HOW THIS BOOK IS ORGANISED

This book starts with an overview of pioneering projects in e-Learning development in Medical Physics and Engineering and continues to provide a well-planned and insightful approach to the entire process of building an e-Learning Unit. This book is based on examples and provides over 60 screenshots aiming to help the beginner in his/her first steps in setting up a course on Moodle, including the essential educational elements.

CHAPTER 1 – e-LEARNING IN MEDICAL PHYSICS AND ENGINEERING: AN OVERVIEW

Starting with a discussion on the definition and types of e-Learning, as well as its advantages for the dynamic professions related to medicine such as Medical Physics and Engineering, this chapter continues with an outline of pioneering and award-winning projects in the profession, accompanied by a detailed description of ways to access them.

CHAPTER 2 – MOODLE AS A VIRTUAL LEARNING ENVIRONMENT (VLE) SYSTEM

This chapter discusses major VLE systems in Higher Education and their typical features and then lists the prerequisites for introducing a VLE System, focusing on the steps the educator needs to undertake to prepare for the seamless introduction of the VLE. The general terms in use in this guide are also defined and a special place is dedicated to copyright on the VLE. The chapters following this one use as an example a course in Medical Imaging Physics and Equipment, which can be a building block of an MSc programme either in Medical Physics or in Medical Engineering (though it is more often associated with Medical Physics). This example can easily be used in any course with image/diagram-rich content.

CHAPTER 3 – BUILDING AN MSC PROGRAMME IN MEDICAL PHYSICS ON MOODLE: THE TEACHER FUNCTIONS IN FOCUS

Here, gaining access to the VLE and log-in are demonstrated and the major roles in the Moodle VLE – Teacher, Manager, and Student – are outlined. The chapter's focus is on the Teacher's VLE role in editing, creating content (in the form of lectures, various types of assessment such as coursework and quizzes), and initiating communication with Students on the VLE (through forums, chats, etc.).

CHAPTER 4 – ROLE-SPECIFIC FUNCTIONS ON MOODLE

While the Teacher's role is mainly related to one building block of the educational programme – the course – the typical role of the Manager is to organise the programme as a whole as it appears on the VLE, in other words to organise and manage the structure of a programme. This is the subject of the first part of Chapter 4. The chapter then continues with illustrating the Student's functions on the VLE, discussing the specific Teacher functions in assessment and finally demonstrating how Moodle can be used as a source of participant information (grades, activity).

CHAPTER 5 – ASPECTS OF MOODLE APPLICATION

This chapter displays results from surveys conducted by the author on e-Learning in Medical Physics and Engineering in lower- and middle-income countries and illustrates ways for building an e-Learning educational programme with a limited budget, comparing/discussing various options.

CLARIFICATION

Throughout this book, the term 'course' is widely used. It corresponds to a building block of an educational Programme on the VLE. In the leading example in the book based on an MSc in Medical Physics, the course is a building block of the MSc programme and, as such, is often delivered in modular form. Thus, in many places in the book, the term 'course' is referred to as 'course in modular form'.

The term 'Manager' is one of the default main roles in Moodle. It should be noted that throughout the book this term does not correspond to a job title and only refers to the role on the VLE. This also applies to the term 'Administrator'.

e-Learning in Medical Physics and Engineering

An Overview

1.1 DEFINITION AND TYPES

The term "e-Learning" evolved with the advancements in the use of technology in education, and it is argued that the prerequisites for its appearance can be traced back to the 1960s and even before [1]. After this exact term was coined in the late 1990s, it came to encompass previously used terms such as "multimedia-based education", "computer-based training (CBT)", and "electronic teaching materials". This term gained popularity after an article in 1998 by A. Morri [2].

It should be noted that there is not a single agreed definition of e-Learning. It is often stressed that e-Learning means different things in different sectors, and nowadays it is generally accepted that in Higher Education, "e-Learning" refers to the use of both software-based and online learning. An important point to note is that e-Learning not only makes available documents in electronic format via online learning, but also encourages interaction between the different users and develops a flexible pedagogical approach. A good source that explores the basic concepts as well as day-to-day issues of e-Learning, looking at both theoretical concepts and practical implementation issues, is Ref. [3].

A variety of accompanying terms are in use in e-Learning–related literature, the most frequently used being Virtual Learning Environment (VLE) and Learning Management System (LMS), and also Managed Learning Environment (MLE), Learning Support System (LSS), etc. All these terms refer to online platforms and software systems configured to facilitate management and student involvement in e-Learning, and describe a wide range of systems that organise and provide access to online education services for students, teachers, and administrators. These services usually include access control, provision of learning content, communication tools, and administration of user groups. Throughout this book, the term "Virtual Learning Environment" (VLE) is accepted to describe the variety of functions mentioned above.

1.2 THE CASE FOR e-LEARNING

Today e-Learning is an intrinsic part of education, and there is hardly any need to make the case for its implementation. We shall discuss briefly here the three main issues driving the development of e-Learning:

- Enhanced quality of learning,

- Increased number of students,

- Facilitating the management of learning.

The evolution of e-Learning in Higher Education has witnessed examples when the driving force for the introduction of e-Learning has been economic efficiency alone, but in our experience and to achieve sustainability, the other factors are discussed here with priority.

1.2.1 Enhanced Quality of Learning

The improved quality of content in e-Learning does not happen automatically. As pointed in [4], this requires not only detailed knowledge of the subject area, but also advanced knowledge of state-of-the-art computer technology and software, and good vision on combining them in a learning pack.

As discussed in a variety of papers, e-Learning is moving more and more from representing an instruction paradigm to becoming a learning paradigm, where the teacher with a learning perspective provides students support and the students themselves actively discover and construct knowledge [5–7].

Facilitating the management of learning is what makes the VLE platform a welcome feature of contemporary education, and this will be illustrated through the example of the Moodle platform throughout this book.

Medical Physics and Medical Engineering are areas that make the case for the introduction of e-Learning a very strong one. It makes possible not only the cost-effective dissemination of a large volume of imaging material, specific for the profession, but it can also accommodate simulations for the explanation and illustration of the scientific fundament and complex functions of various aspects of medical technology. It is also extremely beneficial for such a dynamic profession, allowing for quick and efficient upgrading, and thus making possible the newest updates to reach the targeted student audience in a timely fashion. Pioneering educators in the field, such as the team behind the projects European Medical Radiation Learning Development (EMERALD), European Medical Imaging Technology Training (EMIT), and European Medical Imaging Technology e-Encyclopaedia for Lifelong Learning (EMITEL), discussed further down, developed the first image database in the profession, the first e-Dictionary and e-Encyclopaedia and others, and saw this potential of "computer-aided education" long before the term "e-Learning" was coined.

1.2.2 Increased Number of Students

While not a goal in itself and not the main driving force, this was another achievement and a logical consequence of the pioneering projects mentioned above. Thus the number of students in Medical Physics and Clinical Engineering at the MSc programme in Medical Engineering and Physics (MEP) at King's College London, UK, increased from 12 to 38 in 2011 with the same administrative resources and consequently from 2011 to 2013 reached the number of 112 MSc students (mostly part-time) in three cohorts, becoming one of the largest courses in the profession worldwide. This remarkable economic efficiency was achieved through implementing a pilot Moodle VLE in 2011 at King's College London with retained and even enhanced quality of teaching, which was reflected in the reviews of external assessors over the years. In yet another example, the free e-Learning pack EMERALD, the first such pack to be developed in Medical Physics in 1996, was made available to about 50 students at the ICTP (International Centre for Theoretical Physics) College on Medical Physics, Trieste, Italy and consequently was opened for access free of charge to all medical physicists from low- and middle-income countries, which led to a worldwide use of e-materials (Figure 1.1).

The case for introduction of e-Learning in the profession of Medical Physics and Clinical Engineering becomes even stronger when taking into consideration the predicted needs of tripling the number of medical physicists worldwide by 2035 [8,9].

As this is compared with the growth of the global number of medical physicists in the period of 1965–2015 [10], it is evident that there is an enormous task before the educators in this profession for less than two decades from now. The aim of the current book is to facilitate reaching this target by 2035.

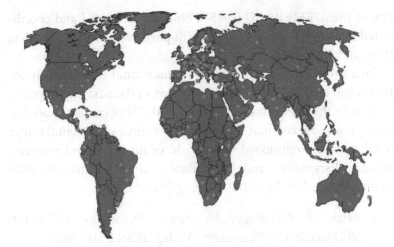

FIGURE 1.1 EMERALD, EMIT, and EMITEL materials disseminated in more than 100 countries.

1.3 PIONEERS OF e-LEARNING IN MEDICAL PHYSICS

As it was pointed before, pioneering educators in the field of Medical Physics realised the potential of education aided by computers long before the term "e-Learning" appeared.

The prerequisites for this were the image-rich contents of the subject and the complexity of the scientific principles involved in the explanation and illustration of the functioning of medical technology.

During the 1990s, it was obvious that the slow paper print of image-rich content did not allow quick educational content update in some dynamic professions. Medical Physics was one of them, and it was obvious that it was necessary to use IT technology in this case.

The first international Conference on Medical Physics Education and Training held in November 1994 in Budapest highlighted the need for international collaboration for the advancement of the profession [11]. Important links between professionals from different countries were forged which formed the basis for future projects in education, supported by the European Union. The first one of them, EMERALD (1995–1998), was managed and coordinated by King's College London with partners from UK, Sweden, Italy, and Portugal [12].

This project developed the first educational databases in the profession on CD-ROM, which were one of the first digital publications in the world to be published with ISBN (as paper books). It was not a surprise that the world's first three such digital image databases were developed in the field of medicine and medical-related disciplines, and remarkably, almost simultaneously (within 4 months), EMERALD being one of them:

- Atlas of Pathology: Urological Pathology CD-ROM, 30 December 1997, Springer-Verlag, ISBN 3540146571

- EMERALD Image Database, Training Courses in Medical Radiation Physics CD-ROM, 19 February 1998, King's College London, ISBN 1870722035

- Developmental Psychology Image Database CD-ROM, 30 April 1998, McGraw-Hill, ISBN 0072896914

The content of the EMERALD image database was soon transferred online, and in 1999, this became the first educational website of the Medical Physics profession.

- EMERALD was the first in a series of projects [EMERALD, EMERALD II (Internet Issue), and EMIT] developed in the period of 1996–2003 under the coordination of the Department of Medical Engineering and Physics at King's College London and King's College Hospital, supported by the EU Leonardo da Vinci programme and bringing together a Consortium of universities and hospitals in the UK, Sweden, Italy, Portugal, Ireland, Northern Ireland, France, Bulgaria, the Czech Republic, and the Abdus Salam ICTP in Trieste. The acronyms stand for European Medical Radiation Learning Development (EMERALD) and European Medical Imaging Technology Training (EMIT). The objectives of

EMERALD and EMERALD II (EMERALD – Internet Issue) were to produce training curricula and e-Learning materials to facilitate core training for medical physicists in three areas (modules): Physics of X-ray Diagnostic Radiology, Nuclear Medicine, and Radiotherapy. EMIT concentrated on the development of similar curricula and e-Learning materials in two further fields: Diagnostic Ultrasound and Magnetic Resonance Imaging. Together the two pilot projects (EMERALD and EMIT) have developed a total of five structured training modules, each having approximately 50 practical training tasks. The detailed curricula (timetables – approximately 10 pages each) of all five modules are available freely (with a volume of approximately 30 MB) at the website www.emerald2.eu.

- The series continued with the multilingual Dictionary in Medical Physics, currently available in 30 languages and 11 alphabets (the Dictionary started in 2003 and continues to be updated). It is available at www.emerald2.eu, and the current supported languages are Arabic, Bengal, Bulgarian, Chinese, Croatian, Czech, English, Estonian, Finnish, French, German, Georgian, Greek, Hungarian, Italian, Japanese, Korean, Latvian, Lithuanian, Malaysian, Persian, Polish, Portuguese, Romanian, Russian, Slovenian, Spanish, Swedish, Thai, and Turkish.

- Almost simultaneously with EMERALD was developed the Sprawls Resources website, www.sprawls.org (2000), and later the IAEA RPOP, www.iaea.org/resources/rpop (2004). These are some of the most visited websites in Medical Physics worldwide and, together with a very useful book on the Internet of Radiology Practice by A. Mehta, published in 2003, are mentioned in the next chapter.

- The EMITEL e-Encyclopaedia, 2005–2009 and with current updates, is the first e-Encyclopaedia in the profession, encompassing more than 3,200 terms in Medical Physics. It is created by a Consortium of the core partners of the projects EMERALD and EMIT, together with the International Network EMITEL (currently 300+ specialists from 36 countries), which continue to care for the EMITEL e-Encyclopaedia and Dictionary (now being updated). The content of the EMITEL e-Encyclopaedia, www.emitel2.eu, is linked to the existing EMERALD and EMIT materials, and it additionally includes Radiation Protection and Hospital Safety terms, thus forming a one-stop knowledge database for those who want to acquire a specific competence and for those who want to refresh their knowledge and learn about the new developments in the profession. The tool is also linked to the multilingual Digital Dictionary, currently including 30 languages and 11 alphabets and providing cross translation between all languages. Further details about the projects EMERALD, EMIT, and EMITEL can be found in [13] and [14]. In December 2004, the European Union recognised the pioneering contribution of the EMIT Consortium to education and e-Learning with the inaugural EU Award "Leonardo da Vinci" (Figure 1.2). The author is one of the contributing members to the Consortium of those projects.

FIGURE 1.2   The Leonardo da Vinci Award Certificate.

The EMITEL Encyclopaedia and the Dictionary are constantly being updated with the concerted effort of over 300 specialists worldwide; it is very popular in the professional community of medical physicists and continues to be visited by over 8,000 specialists per month from all over the world (Figures 1.3 and 1.4), while the visitors in a typical single day amount to over 300 with over four page impressions per visitor, which excludes occasional visitors/error clicks (Figures 1.5 and 1.6).

Similar official statistics are available for the EMERALD materials (Figure 1.7).

FIGURE 1.3 EMITEL visitors over a 30-day period in May 2018; illustration. (Source: Official 1&1 server statistics.)

FIGURE 1.4 Visitor location of EMITEL over a 30-day period in May 2018; illustration. (Source: Official 1&1 server statistics.)

FIGURE 1.5 EMITEL visitors in one day (a typical day in May 2018) with a per-hour graph; illustration. (Source: Official 1&1 server statistics.)

FIGURE 1.6 Page impressions of EMITEL visitors in one day per hour (a typical day in May 2018); illustration. (Source: Official 1&1 server statistics.)

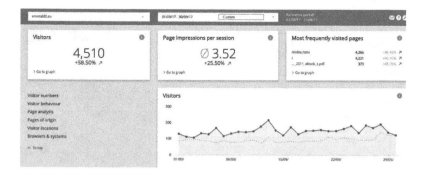

FIGURE 1.7 EMERALD visitors over a 30-day period in 2017; illustration. (Source: Official 1&1 server statistics.)

The trend steadily continued in 2019, and the worldwide popularity of the materials is illustrated in Figure 1.8.

The popularity of those projects among the professional community triggered response at global fora such as the high-level UNESCO World Conference "Physics and Sustainable Development" (Durban, South Africa, 2005), where specialists from all over the world discussed the future trends in applied physics. All e-Learning materials EMERALD/EMIT and the Sprawls Foundation ones (see Chapter 1.4) were presented at the forum and attracted widespread interest. This supported the UNESCO Conference resolution to highlight the topic "Physics and Health" as one of just four specific topics for the development of applied physics in the 21st century.

The widespread global use of the EMITEL e-Encyclopaedia with Dictionary together with the earlier materials of EMERALD and EMIT resulted in a number of conferences, workshops, and other activities in the field of Medical Physics, discussing specific professional and educational topics. This led to the creation of an international forum on these issues in the form of a journal. Thus, the journal *"Medical Physics International"* was launched at the end of 2012. This is a free online journal of the

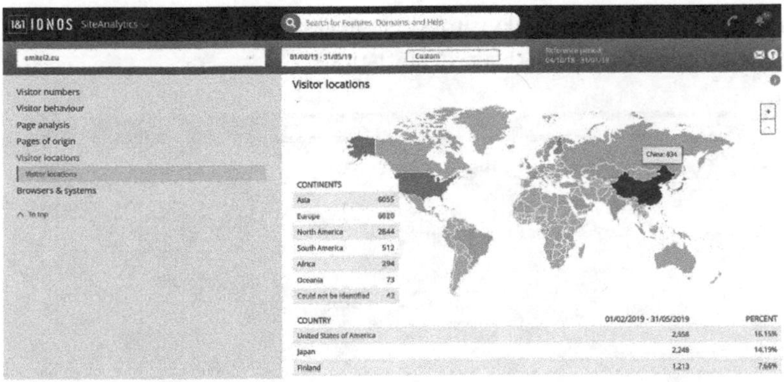

FIGURE 1.8 Statistics on EMITEL visitors per continent over 4 months in 2019. (Source: Official 1&1 server statistics.)

International Organisation for Medical Physics (IOMP) with co-editors P. Sprawls and S. Tabakov, which can be accessed at www.mpijournal.org. Since its first issue in April 2013, the journal attracts tens of thousands of professionals each month (Figure 1.9), and this trend continues steadily to date (Figure 1.10).

Among its professional and educational aims, the journal contributes to the dissemination of information about e-Learning materials, which normally have a short life and thus facilitates the much-needed quick access to those materials for professionals worldwide.

Further information about all the listed projects is available in [15], and free downloadable educational materials can be accessed on www.emerald2.eu.

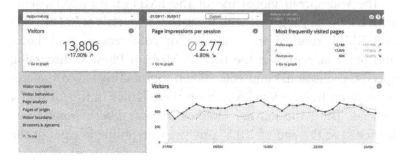

FIGURE 1.9   MPI Journal visitors over a 30-day period in 2017; illustration. (Source: Official 1&1 server statistics.)

FIGURE 1.10   MPI Journal statistics of visitors per continent, 5 months in 2019. (Source: Official 1&1 server statistics.)

1.4 OTHER LEADING e-LEARNING PROJECTS IN MEDICAL PHYSICS, ENGINEERING, AND RELATED DISCIPLINES

The only other electronic teaching materials in medicine, known to the EMERALD Consortium at the beginning of that project (1995), were in King's College London, UK (developed by N. J. D. Smith in 1992) – this was the Distance Learning course on Dental Radiography, using images on *Photo CD* – the new (at that time) invention of *KODAK*, using a special CD player (*Photo CD player* or *CD-i player*). The course was running successfully for several years before the rapid expansion of CD-ROM technology, which replaced the photo CD and triggered a pioneering web-based distance learning course in Dentistry in the beginning of the 2000s, in which the author also participated actively from 2003 to 2008.

At the time of the projects EMERALD II and EMIT, a number of other Medical Physics and related e-Learning activities were initiated around the world. In fact, the term "e-Learning" gained popularity in 1998 through the paper "A Bright Future for Distance Learning: One Touch/High Alliance Promotes Interactive 'e-Learning' service" [2].

Almost in parallel with EMERALD, other websites were also developed.

One of the first and also a very popular one is the website of the Sprawls Educational Foundation, dedicated to enhancing the learning and teaching of Medical Imaging Physics, Engineering, and Technology, and the utilization of technology to enhance human performance in the educational process. It contains the *Physical Principles of Medical Imaging – Online*, a free and open resource for learning and teaching the physics and technology of medical imaging. It provides high-quality visuals to be used by onsite learning facilitators, web-based visuals formatted for tele-teaching, online modules for use in hybrid courses or self-study, teaching/study guides, and learning objectives [16].

The first paper on tele-teaching in Medical Physics in 1999 covers important aspects of the educational paradigm [17] and a book including an extensive list of websites with e-Learning materials related primarily to radiology, but also very useful for Medical Physics, was published by Springer-Verlag in 2003 [18].

At the same time, there was an important development of software aimed at creating simulations – a very important element of a profession applying radiation to create the medical images. Some of the first such simulators were the X-Ray equipment interactive simulators created at the University of Patras [19] and in Università degli Studi di Cagliari [20].

Among the early developments in this field can be listed also the Gamma Camera DOS-simulator learning pack, the PowerLab Systems of AD Instruments, various LabView simulations, and the IPEM X-ray Spectrum Processor software, and more recent effective simulations include the VERT (Virtual Environment for Radiotherapy) developed by the University of Liverpool [21]. The importance of e-Learning prompted the *Journal Medical Engineering and Physics* to dedicate a special issue on the subject in 2005. Its editorial describes briefly existing simulations and other e-Learning projects in the profession at the time [4]. Yet further examples of simulations can be found at the websites of various manufacturers and leading professional societies such as AAPM and IPEM.

Simulations are a very effective teaching tool, but several considerations have to be taken into account: their creation requires a combination of excellent IT skills and good pedagogical approach; they are also very time-consuming. On the other hand, a specific problem with simulations is their relatively short life cycle (often less than 5 years). The main reason for this is the upgrade of software platforms, which in most cases has restricted the use of simulations.

Some other open and free sources are listed below.

The scientific journals published by professional societies such as AAPM, ACMP, RSNA, and many others are significant resources for continuing education and professional development.

1.4.1 Training Websites

- The AAPM Virtual Library provides online access to selected continuing education courses presented at the annual meetings. The video format includes both the visuals and the presenters' vocal discussion.

These are available to members, and many of the AAPM Educational Resources are also available to medical physicists in the developing countries. Access requires registration as a Developing Country Educational Associate (DCEA) and obtaining a 'Developing Country Educational Associate' USERNAME and PASSWORD.

- AMPLE – or the Advanced Medical Physics Learning Environment, an IAEA-developed e-Learning tool used in Asia and the Pacific to help support Medical Physics training in the fields of treatment planning and ensuring that diagnostic imaging provides optimal results with minimal radiation exposure to the patient.

- RPOP – Radiation Protection of Patients, a website by IAEA and one of the most popular sources for patients and public on the safe and effective use of radiation in medicine; www. iaea.org/resources/rpop

- Various courses for online training provided by IAEA and periodically announced at www.iaea.org/about/organizational-structure/department-of-nuclear-sciences-and-applications/division-of-human-health/dosimetry-and-medical-radiation-physics-section

1.5 MANAGING CONTENT (IN-HOUSE, PROPRIETARY, OPEN SOURCE)

The development of platforms for e-Learning management became rigorous since the late 1990s. In Medical Physics and Engineering, there were successful examples for creating in-house platforms such as the ERM e-Learning system for education in

Medical Radiation Physics [22] and the KISS e-Learning system in Medical Engineering [23]. Both papers describe the technical requirements (server, browser) in much detail and provide useful insights into the architecture of e-Learning systems. At approximately this period, a very successful virtual campus for Biomedical Engineering (BME) as well was developed in Finland [24]. However, such decisions require a dedicated IT team for development and support, and the period for the development of such a system can be prohibitively long.

Parallel to that appeared proprietary e-Learning VLE systems, which started being licensed to education institutions, the most popular of which were WebCT and its successor, Blackboard Learning Management System (currently Blackboard Learn™). They quickly became more popular than the in-house system, as they offered a shell that could be adapted to different educational programmes within an institution and saved time and effort for their initial creation.

Since the 2000s, successful courses in the fields of Medical Physics were developed using those two VLE systems. Examples include the following:

- For Blackboard:

    - a research program in Medical Physics at the University of Adelaide, Australia [25]

    - a Postgraduate Nuclear Medicine Imaging in Salford University, UK [26]

    - an MSc in Biomedical Engineering at Imperial College, London, UK [27]

    - an MSc in Medical Physics at Newcastle University, UK [28]

- For WebCT

    - WebCT – a Diagnostic Ultrasound Course at Monash University Melbourne, Australia [29]

- a course on Nuclear Medicine Technology at the University of Arkansas for Medical Sciences, US [30]

A major disadvantage for institutions with limited resources is the fact that both WebCT and Blackboard use proprietary software (also known as "closed-source software") because the software developer retains the rights over the source code, and they are available for use under a paid fee for licensing.

In the late 1990s, the Moodle VLE was developed. It has continued to evolve since 1999 (since 2001 with the current architecture). Its most important feature is being a free and open-source LMS with customizable management features. It is now one of the most widely used VLE systems in academia [31,32].

Institutions can add as many Moodle servers as needed without having to pay license fees, and the open-source software means developers can make modifications based on their needs. Based on these features, the Moodle platform is used to develop the examples of e-Learning courses in Medical Physics and Engineering in this book. It is worthwhile to note here that many leading universities such as University College London, King's College London, and the Open University, UK have adopted Moodle as their e-Learning platform; in the case of King's College London, the flexibility of the open source has been used to create the modification KEATS (King's E-learning and Teaching Service) based on the specific requirements of the university.

BIBLIOGRAPHY

1. The History of eLearning (infographic), EdTechTimes staff, Dec 17, 2012; https://edtechtimes.com/2012/12/17/the-history-of-elearning-infographic/, last accessed 21 Nov 2019.
2. Morri, A. "A Bright Future for Distance Learning: One Touch/High Alliance Promotes Interactive 'e-Learning' Service", *Connected Planet*, Nov 1997.
3. Masters, K. and Ellaway, R. e-Learning in medical education Guide 32 Part 1: learning, teaching and assessment. *Medical Teacher* 2008; 30: 455–473.

4. Tabakov, S. e-Learning in medical engineering and physics, Guest editorial. *Medical Engineering & Physics*, Elsevier, 2005; 27 (7): 543–547.

5. Barr, R. B. and Tagg, J. From teaching to learning – a new paradigm for undergraduate education. *Change* 1995; 27: 12–26.

6. Alexander, S. E-learning developments and experiences. *Education and Training* 2001; 43: 240–248.

7. Stephenson, J. (ed.). *Teaching and Learning Online Pedagogies for New Technologies*, London: Kogan Page, 2001.

8. Rifat Atun, D., Jaffray, A., Barton, M. B., Bray, F., Baumann, M., Vikram, B., Hanna, T. P., Knaul, F. M., Lievens, Y., Lui, T. Y. M., Milosevic, M., O'Sullivan, B., Rodin, D. L., Rosenblatt, E., Van Dyk, J., Yap, M. L., Zubizarreta, E., and Gospodarowicz, M. (2015). Expanding global access to radiotherapy. *Lancet Oncology* 2015; 16: 1153–1186.

9. Tsapaki, V., Tabakov, S., and Rehani, M., Medical Physics workforce: a global perspective. *European Journal of Medical Physics* 2018; 55: 33–39.

10. Tabakov, S. Global number of medical physicists and its growth 1965–2015, *Journal Medical Physics International* 2016; 4(2): 79–81.

11. Roberts, C., Tabakov, S., and Lewis, C. (eds). *Medical Radiation Physics: A European Perspective*, London: King's College London, 1995, available from: www.emerald2.eu/mep/e-book95/MedRadPhys_95b.pdf.

12. Strand, S., Milano, F., Tabakov, S., Roberts, C., Lamm, I., Liungberg, M., Benini, A., daSilva, G., Jonson, B., Jonson, L., Lewis, C., Teixeira, N., Compagnucci, A., and Ricardi, L. EMERALD for vocational training and interactive learning, 18 symposium on radioactive isotopes in clinical medicine. *European Journal of Nuclear Medicine*, Springer, 1998; 25 (1): 1–6.

13. Tabakov, S., Roberts, C., Jonsson, B., Ljungberg, M., Lewis, C., Strand, S., Lamm, I., Milano, F., Wirestam, R., Simmons, A., Deane, C., Goss, D., Aitken, V., Noel, A., and Giraud, J. Development of Educational Image Databases and e-books for Medical Physics training. *Journal Medical Engineering and Physics*, Elsevier, 2005; 27 (7): 591–599.

14. Tabakov, S., Smith, P., Milano, F., Strand, S-E., Lamm, I-L., Lewis, C., Stoeva, M., Cvetkov, A., and Tabakova, V. EMITEL e-Encyclopaedia of Medical Physics with Multilingual Dictionary

in Medical Physics and Engineering Education and Training – Part I, ICTP, Trieste, Italy, 2011, 234–240, ISBN 92-95003-44-6, available from: www.emerald2.eu/mep_11.html.

15. Tabakov, S. and Tabakova, V. *The Pioneering of e-Learning in Medical Physics: The Development of e-Books, Image Databases, Dictionary and Encyclopaedia*, London: Valonius Press, 2015, ISBN 978-0-9552108-3-9.

16. www.sprawls.org/resources, last accessed 23 Nov 2019.

17. Sprawls, P. and Ng, K. H. Teleteaching Medical Physics. 1999 AAPM annual meeting, Nashville. *Medical Physics* 1999; 26 (6): 1055.

18. Mehta, A. *The Internet for Radiology Practice*, New York: Springer, 2003, ISBN 0-387-95172-5.

19. Palikarakis, N. Development and evaluation of an ODL course on Medical Image Processing. *Medical Engineering & Physics* 2005; 27(7): 549–554.

20. Fanti, V., Marzeddu, R., Massazza, G., and Randaccio, P. A simulation tool to support teaching and learning the operation of X-ray imaging systems. *Medical Engineering & Physics* 2005; 27 (7): 555–559.

21. Kirby, M. Virtual environment for teaching Radiotherapy Physics – a four year experience; 16th Learning and Teaching Conference, University of Liverpool, Liverpool, UK 5–6 July 2018, available from: www.researchgate.net/publication/326941826_VERT_and_radiotherapy_Physicstechnology, last accessed 24 Nov 2019.

22. Stoeva, M. and Cvetkov, A. e-Learning System ERM for Medical Radiation Physics Education. *Medical Engineering & Physics* 2005; 27 (7): 605–609.

23. Hutten, H., Stiegmaier, W., and Rauchegger, G. KISS – a new approach to self controlled e-learning of selected chapters in Medical Engineering and other fields at bachelor and master course level. *Medical Engineering & Physics* 2005; 27 (7): 611–616.

24. Kybartaite, A. *Impact of Modern Educational Technologies on Learning Outcomes: Application for e-Learning in Biomedical Engineering*, Tampere: Tampere University of Technology, 2010, ISBN 978-952-15-2385-4; ISSN 1459-2045.

25. Pollard, J. A research program in Medical Physics for remote students. *Medical Engineering & Physics* 2005; 27 (7): 599–603, available from: www.ncbi.nlm.nih.gov/pubmed/16055365.

26. https://beta.salford.ac.uk/courses/postgraduate/nuclear-medicine-imaging.

27. www.imperial.ac.uk/admin-services/ict/self-service/
teaching-learning/elearning-services/blackboard/.
28. www.ncl.ac.uk/media/wwwnclacuk/facultyofmedicalsciences/
files/Medical%20Physics%20Programme%20Handbook%20
1819%20FINAL.pdf.
29. https://core.ac.uk/download/pdf/38301619.pdf.
30. http://citeseerx.ist.psu.edu/viewdoc/download?doi=10.1.1.468.918
&rep=rep1&type=pdf.
31. The Top 8 Open Source Learning Management Systems,
Christopher Pappas, available from: https://elearningindustry.
com/top-open-source-learning-management-systems.
32. Academic LMS Market Share: A view across four global regions,
available from: https://eliterate.us/academic-lms-market-share-view-
across-four-global-regions/.

CHAPTER 2

Moodle as a Virtual Learning Environment (VLE) System

2.1 MAJOR VLE SYSTEMS IN HIGHER EDUCATION: TYPICAL FEATURES

The most popular e-Learning platforms in Higher Education since the mid-1990s are as follows:

- WebCT, developed in 1996 by the University of British Columbia.

- Blackboard, developed in 1997 at Cornell University in Washington.

- Moodle, which started its development in 1999 and exists in its present form since 2001, developed in Perth, Australia by Martin Dougiamas.

It is not surprising that they command about 80% of the total market overall educational share in the world and 87% in academia. Worldwide, Moodle has over 50% of market share in academia in Europe, Latin America, and Oceania, and taking into account the big US Higher Education market, Moodle and Blackboard share the leading position across the globe [1].

Other VLE systems entering rapidly in the education sector are Canvas and D2L.

It is interesting to note that many of the popular and quickly expanding new VLE systems have been developed in Australia. This is certainly due to the fact that Australia has a long history of distance education, and while not the first one to develop this trend, it is the country that has arguably the greatest achievements in providing distance education for wide and varying strata of its population, employing a range of technical inventions both established and innovative ones at the time, such as airplanes for distributing pupils' coursework, the radio and TV for communication [2].

Typical features of the VLE in Higher Education are as follows.

2.1.1 The Possibility to Manage Courses, Users, and Roles

The VLE platform is typically used to create professional structured course content. There are basically three major roles – Teacher, Student, and Manager – and several subsidiary ones, and they are subject to hierarchical relations (i.e., control of access to content for the different roles, tracking student progress, managing and monitoring student attendance and participation).

2.1.2 Online Assessment

The VLE platforms are suitable for both summative and formative assessments. These platforms allow the creation of various types such as offline tasks (i.e., in the form of essays) as well as different multiple question types such as one/multi-line answer, multiple-choice answer, drag-and-drop order, essay, true or false/yes or no, fill in the gaps, and agreement scale.

2.1.3 User Feedback

The VLE platform is a very convenient medium for students' exchange of feedback both with teachers and their peers through discussion groups, forums, etc.

2.2 PREREQUISITES FOR INTRODUCING A VLE SYSTEM

There are a number of prerequisites for the successful adoption of a VLE system in an organisation. One of the major ones is technological availability; in general, it is crucial to secure internet and access of students to computers. We can recognise that this is quite often a problem in lower- and middle-income (LMI) countries, and these will be reviewed in a separate chapter later. Financial considerations also come into the equation at this stage. Then comes the degree of dedication of the educators immediately involved in the process. The slow implementation (adoption rate) of the VLE is due to the combined lack of commitment on behalf of the institution's top management and of those employees who are directly involved in the process [3].

A model that appears to be most useful for post-adoption analysis of an institutional innovation, with the potential for pre-adoption guidance of future practice, is the RIPPLES (Resources, Infrastructure, People, Policies, Learning, Evaluation, and Support) model developed by Surry, Ensminger, and Haab (2005) [4]. This model comprehensively covers a range of factors for consideration including:

- the fiscal *resources* associated with innovation adoption;

- the institution's *infrastructure*, namely, hardware, software, facilities, and network capabilities in support of teaching resources, production resources, communication resources, student resources, and administrative resources;

- needs, hopes, values, skills, and experiences of the *people* involved;

- institutional *policies* and procedures;
- the relationship between the technology and *learning* outcomes;
- *evaluation* and review (both summative and ongoing), including the impact of the technology on learning goals;
- the *support* systems and scaffolding required to ensure successful implementation.

However, Benson and Palaskas [5] discuss another very important initial hurdle in front of adopting a VLE-non-intuitive courseware creation. This coincides directly with the experience of the author when she was working on pilot e-Learning projects since the 1990s. The good quality of the original teaching material and its subsequent preparation for the e-Learning system is, in our opinion, by far one of the most important factors for successful adoption of the VLE. But this has also to be supported by good understanding on behalf of the educators for the principles of building a VLE. As Tabakov (2005) [6] points out, there are many examples when the professional educator and the pedagogist place such unrealistic requirements in front of the IT specialist that the final product becomes not only clumsy and less-effective, but also more expensive. Furthermore, it has to be realistically assessed whether functionality is compromised for the sake of attractive/flashy appearance. The same refers to the introduction of simulations, which are time-consuming to create but more than any other component in the e-Learning media simulations depend on the developing and reading software and often after two to three changes of the software version could lose their initial functionality, if not becoming completely non-readable and/or non-functional. Overall, the successful adoption of a VLE would ensure that the instructional paradigm of classroom learning is enhanced by the learning paradigm, empowering both student and educator (for further discussion of the instructional vs learning paradigm, see Ref. [7]).

The importance on emphasising knowledge structuring prior to adopting the VLE is also supported by the experience of educators developing an internet-based course in Ref. [8].

As this stage is considered crucial by the author for the successful introduction of an e-Learning platform, Section 2.4 is dedicated to the first steps the educator needs to undertake so that the VLE is adopted seamlessly after defining the general terms in use in this guide.

2.3 GENERAL TERMS IN USE IN THE GUIDE TO BUILDING OF A MOODLE COURSE

- Course – this term throughout the book corresponds to a building block of the MSc programme. In the examples to follow in the next chapters, the course is in modular form and could also be referred to as a 'course in modular form'.

- The person who creates the space available for the whole Programme in the book is referred to as 'space creator'. Within an educational institution developing an e-Learning platform for many different programmes this is normally a person with specialist IT knowledge who works within an IT team/department. In such case, the 'space creator' will have limited role in the management of the MSc programme once the space for it is available on the VLE. On the other hand, if the space for the programme is created independently and with limited IT support (illustrated in Chapter 5), the person creating it will also participate actively in managing the site.

- Programme Director – the person responsible for the overall academic delivery of the MSc programme.

- Manager – this is one of the default main roles in Moodle. It is in effect the e-Learning Manager. Wherever the term "Manager" appears within the book, it is used in the sense of the role of 'Manager' within Moodle (while in educational institutions, the same person's job title may be different).

- Teacher – this is another of the default roles in Moodle. The functions of the Teacher are mainly linked to the preparation of the content of the course and student assessment.

The three main roles – Manager, Teacher, and Student – as well as some auxiliary roles such as non-editing Teacher are discussed in detail in Chapter 3 with examples of what functions they perform within the VLE.

- Administrator – in the book this role appears in Chapter 5 where the creation of the Programme space is performed on the Moodle server – MoodleCloud. This is a role which assumes higher responsibilities than the e-Learning Manager as s/he is also a site-creator. Normally what would differentiate this Administrator creating a space on MoodleCloud, from the 'space' creator in case of an institutional platform is the level of knowledge required to build the site from scratch. In the case of the Administrator building a site on MoodleCloud and his/her role as described in Chapter 5, the necessary specialised IT knowledge is considerably less than this of a 'Space' Creator belonging to an IT department. On the other hand, it is helpful (although not strictly obligatory) for the Administrator to have first-hand knowledge of the subjects covered in the educational programme delivered on the VLE.

The examples in the following chapters are of an MSc in Medical Physics for clarity and consistency, but they contain general principles, valid not only for other levels of education in this discipline, but also to other medical and medical-related specialties as well as many disciplines in the Humanities and everywhere where there is a strong imaging component.

2.4 THE MOODLE VLE: TEACHER PREPARATION

First and foremost, in our experience the lecture material has to be well structured and the supporting sources to be gathered and clearly arranged.

It is customary nowadays for all lecturers to supply their lecture material in Power Point (PPT) format which will then be converted to handouts in PDF format. Often development of diagrams, formulas, etc. for educational purposes requires additional software and is time consuming. The use of an Interactive White Board (IWB) during the test run of the lectures can facilitate this. With the IWB the lecturer could draw/write by hand the diagrams/formulas directly over the pre-prepared PPT slides (or add extra slides in the existing sequence). The summative result is saved in the PPT file and later transferred on PDF handouts.

In specific cases, lecture material can be directly in text format (Word/Adobe PDF) – and then very often standard fonts are used such as *Times New Roman* or *Arial*. The images could be customarily used with normal screen resolution (usually 72 dpi) and for specialised images higher resolution will be required.

In all cases, it must be assured that the files uploaded on the VLE system are compatible for different computer systems which the students might use. Word and Adobe PDF will meet these requirements. Note-taking software (e.g., Evernote on a tablet) may be recommended to students.

There follows the preparation of video/audio supporting materials: these should be in optimum file size, and any issues should be discussed with the VLE Manager for the programme.

Additional materials such as URLs, e-books, and e-Journals should be prepared in advance. Copyright should be checked at a very early stage of preparation.

Because of their importance, copyright checks and preparations are discussed in more detail in Section 2.5 and are also mentioned in Chapter 3, Section 3.4 which deals with creating content on the VLE. In Section 3.4, the educator can find also recommendations for the formatting of the content to upload on the VLE and for formatting of the whole course.

In brief the preparation of the Teacher to transfer the lecture material on the VLE should be summarised in the following points:

- Lecture structuring.

- PPT handouts or Word/Adobe text files. Four slides on a landscape – formatted page, for the handouts is a very successful formatting, in some cases a formatting of two slides per page, portrait format is more suitable, i.e., if there are images with important details. An example of this formatting with an image is available in Chapter 3.

- Formatting of the content – such as font (recommended san serif), size, and format of the images to be adapted; for normal images, JPEG format with 72 dpi; for specialised images, higher resolution.

- Preparation of audio/video material – this stage should be discussed with e-Manager for size recommendations.

- Supplementary materials – i.e., links to internet websites, e-books, and e-journals, etc.

- Copyright checks and steps of securing copyright permission (this can be time consuming, so it is necessary to plan well ahead); please see also Section 2.5.

- Possibilities for personalisation of the appearance, layout of the course – this stage should be checked with Programme Director and e-Manager of the course.

- Preparation of coursework tasks, quizzes (including a system of assessment).

In principle, the Teacher is not alone in developing the course on the Platform, and he/she is always supported by the Manager of the course on the VLE.

- Personalising the module, linked with its customisation – this is a step that has to be discussed between the Leader of the course and the VLE Manager taking into consideration factors such as available space and coordination with other modules.

- Decision on formatting of the course according to the requirements of the programme curriculum. Here the decision is between topic/weekly formatting. A weekly formatting will be convenient/useful if a particular subject from the curriculum is delivered throughout a given term, i.e., 2 h weekly. This is typical of a BSc programme, in sciences in general. For instance, a BSc programme in Biomedical Engineering in its first term would typically include Fundamentals of Mathematics and Physics and other core subjects related to the disciplines [9,10] and the weekly programme for delivering the lectures will consist of a mixture of couple of hours of those core subjects; therefore, it is suitable to select a weekly format for the layout on the VLE.

In contrast, an MSc Medical Physics is delivered typically in modular form [11] mainly due to the availability of specialist lecturers. Therefore, the subjects in the curriculum will be developed on the VLE in topics format. The two formats are discussed further in Chapter 3.

- Another important step for implementing the VLE is preparation of the coursework and other assessment. Before preparing the content of coursework and assessment, a first step is to decide on the system of assessment and weight of the different ingredients. It is necessary at this stage to distinguish between summative and formative assessments and decide which of the formative ones will be included in the final grade for the programme, in close consultation with the programme specifications.

- As a conclusion, Teachers should bear in mind that a course on the VLE is built in close collaboration with the VLE Manager, thus preventing major pitfalls in the future and securing smooth transition from paper to VLE.

2.5 COPYRIGHT AND THE VLE PLATFORM

The question of copyright is especially important in e-Learning as the materials are made available on the internet.

Although this does not seem to be any different from securing copyright clearance for non-electronic format lectures, in the internet environment there is an additional dimension because reproducing (and distributing) handouts is facilitated considerably in an e-format – by a single click for download and then forwarding the file as an e-mail attachment for example. The task for conforming to copyright requirements can be facilitated to a great extent if the following steps are taken:

- The initial step in this direction should be checking if the copyrighted material is covered by a university-broad licence. This is often the case in many countries; for instance, in the UK this is managed by the Copyright Licensing Agency (CLA), and a CLA university licence secures a blanket clearance to copies (within limits) of the content the institution members need.

- If the intended content is not covered by such type of licence, it is necessary to seek permission from the copyright owner. Things may be complicated by the fact that the author is not necessarily the copyright owner (it may be their publisher, etc.). It is, however, worth contacting the author/s directly to start locating the copyright owner. Another source of such information is the legal department of the university/company where the author(s) is(are) employed. When addressing the copyright owner seeking permission to use the copyrighted material, a very important consideration that needs to be highlighted in the correspondence is that the material would be available for limited access only, to a limited number of users on a secure platform (and not freely with unlimited access on the internet). It is courteous also to discuss from the start the type of acknowledgement (i.e., the figure/diagram is reproduced by permission of.../courtesy of...).

BIBLIOGRAPHY

1. Academic LMS Market Share: A View across Four Global Regions, available from: https://eliterate.us/academic-lms-market-share-view-across-four-global-regions/
2. Stacey, E. The history of distance education in Australia. *Quarterly Review of Distance Education* 2005; 6 (3): 253–259.
3. Mallak, L. Challenges in Implementing e-Learning, available from: http://scholar.google.co.uk/scholar_url?url=https://www.researchgate.net/profile/Larry_Mallak/publication/224073836_Challenges_in_implementing_e-learning/links/553127920cf27acb0dea4ced/Challenges-in-implementing-e-learning&hl=en&sa=X&scisig=AAGBfm1_Jk_RLJyVeYoYbAeAiiNbVU_Iqg&nossl=1&oi=scholarr, last accessed 27 Nov 2019.
4. Surry, D. W., Ensminger, D. C., and Haab, M. A model for integrating instructional technology into higher education. *British Journal of Education Technology* 2005; 36: 327–329, available from: http://dx.doi.org/10.1111/j.1467-8535.2005.00461.x, last accessed 27 Nov 2019.
5. Benson, R. L. and Palaskas, T. Introducing a new learning management system: an institutional case study. *Australasian Journal of Educational Technology* 2006; 22 (4): 548–567.
6. Tabakov, S. e-Learning in medical engineering and physics, Guest editorial. *Medical Engineering & Physics*, Elsevier, 2005; 27 (7): 543–547.
7. Shifting Paradigms: Moving from an Instructional Paradigm to a Learning Paradigm November 28, 2017 by Institute for Personalized Learning, available from: www.competency-works.org/uncategorized/shifting-paradigms-moving-from-an-instructional-paradigm-to-a-learning-paradigm/.
8. Jönsson, B. A. Medical Physics for secondary-level teachers. *Medical Engineering & Physics* 2005; 27 (7): 571–581. A case study

of successful e-learning: a web-based distance course in medical physics held for school teachers of the upper secondary level.

9. Physics with Medical Physics, available from: www.ucl.ac.uk/ medical-physics-biomedical-engineering/study/undergraduate/ physics-medical-physics-bsc.

10. Physics with Medical Physics, available from: www.cardiff.ac.uk/ study/undergraduate/courses/2020/physics-with-medical-physics-bsc.

11. Tabakov, S., Sprawls, P., Krisanachinda, A., Podgorsak, E., and Lewis, C. IOMP model curriculum for post-graduate (MSc-Level) education programme on Medical Physics. *Journal Medical Physics International* 2013; 1: 16–22, available from: www.mpijournal.org/ pdf/2013-01/MPI-2013-01-p015.pdf.

CHAPTER 3

Building an MSc Programme in Medical Physics on Moodle

The Teacher Functions in Focus

3.1 ACCESS TO THE VIRTUAL LEARNING ENVIRONMENT (VLE)

In this chapter, we shall start with a discussion of a scenario where the space for the MSc programme on the Moodle VLE has been created beforehand. In an educational institution, this 'space creator' will most probably be a dedicated person from the IT department.

Once this person sets up the required space, s/he would rarely interfere with the process of managing the content inside and that would typically be only after a request from the Manager of the course or another authorised person.

There is also a way to create this space independently, which is illustrated in Chapter 5. There it will be illustrated how to proceed if there is no dedicated IT person for this initial stage and none or very limited IT support for the programme. In that case the 'space creator' will also participate most actively in managing the site, creating and organising its content. For the purposes of this book the author has created a dedicated space for a sample of an education Programme on Medical Physics on Moodle called 'EMERALD Medical Physics' using this independent approach.

As the starting point for our current scenario in this Chapter 3 is the already created space, the entry point is the log-in page, requiring as usual a username and a password. (Figure 3.1) In both cases above, a link to the VLE and an initial password will be communicated by the 'space creator' to a new user in advance, usually with the requirement to change it to a more convenient one after the first log-in.

In most cases when the education institution creates the space for the education programme on the VLE, full access to the space

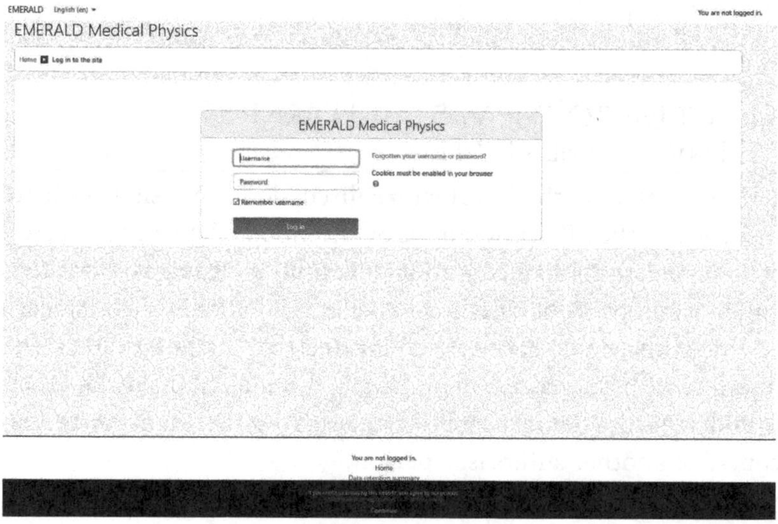

FIGURE 3.1   Log-in page.

will be given to own staff involved in that programme. (Those people would already have an e-mail account at the institution.) Full access can also be granted to external users depending on their functions on the programme, for instance for visiting lecturers, who need to upload their lectures and/or grade student coursework. Granting full access in such cases can be achieved by setting up an affiliated e-mail account at the host institution. For other participants like external examiners, external programme assessors, and accreditation bodies, setting up an affiliated account will not be necessary normally, as they will not be expected to alter/contribute to the content of the platform. In such case, a Guest account would be more appropriate and that would again be taken care of by the IT department.

The example discussed in this chapter is based on the assumption that the building blocks of the educational programme, the courses, are delivered in a modular basis, intensively over a short period of time (i.e., a week). This is most often the case in current MSc programmes in Medical Physics in various universities as many of the courses are highly specialised and would depend on the availability of the lecturers, who are normally professionals who combine academic with clinical experience. This modularity is in line with the recommendations of the IOMP Model Curriculum [1] and the IAEA Training Course Series (TCS) 56 [2]. It is also illustrated how this example can be adapted to a course delivered over a longer period of time (i.e., a whole term).

3.2 MAIN PARTICIPANTS ON THE VLE

There are several roles within Moodle, which relate to different access levels for the participants in the Moodle platform varying in functionality, resource access, and management.

The standard names of the three major and most frequently established roles are Manager, Teacher, and Student. (Note: these are default names, but as Moodle is very flexible for personalisation, one can encounter variation of these names in various education institutions if they have been customised at the institution level.)

Auxiliary or derivative roles such as Non-editing Teacher, useful for a larger course, with a high number of students, will be discussed as well at some point. Throughout the discussion in this chapter and the following one, it will be aimed to illustrate what the functions are for each of these roles.

As discussed earlier, there will be someone responsible for creating the space for the programme, and if this person is from the dedicated IT team at an educational institution, s/he would have no further role in the management of this space, unless for some troubleshooting in case of a major emergency that cannot be solved by other participants.

The role of the Manager will be normally to create the structure of the allocated space, after setting up the front page of a given course on the VLE. The role of the Manager can be performed by non-academic staff in the department delivering the educational programme with an input from the Programme Director.

3.3 COURSE FRONT PAGE AS PRESENTED TO THE TEACHER

We shall start first with the role of the Teacher, taking into account that the internal space of the course has already been struc-tured by the Manager. (The Manager's functions are described in Chapter 4.) The Teacher's role is one that will normally be assigned to a lecturer in a specific subject (in our case, in the area of Medical Physics). As such, after log-in the lecturer will be presented with a screen similar to the one shown in Figure 3.2, which is the front page of the course of Imaging with Ionising Radiation ('IIR Medical Imaging Physics and Equipment'), one of the courses in the Model Curriculum of an MSc in Medical Engineering and Physics MEP [1].

Two of the main building blocks are a central one with five visible sections – Summary, Lectures, Quizzes, Coursework, and Past Exam Papers – and a block to the left, named 'Navigation', which appears with its drop-down menu. The two blocks in Administration – Site administration and Course

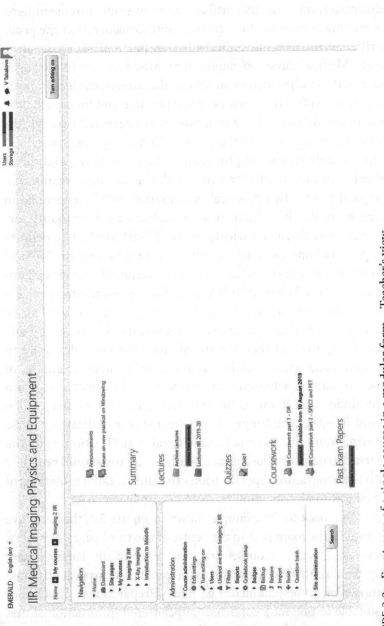

FIGURE 3.2 Front page of a single course in a modular form – Teacher's view.

administration – are also visible. In the example discussed here, a structure is presented which takes into consideration the peculiarities of the area of Medical Physics, particularly at the stage where Medical Physics is taught at an MSc level. In an advanced stage of the studies such as an MSc in the subject, the lecturers are very often professionals with clinical practice and their participation in the delivery of the programme is organised through the modular arrangement of the courses. Thus, a typical specialised course is delivered within the framework of 5–8 days, and this is reflected in the Moodle structure with the sections organised in a topical frame. This practical organisation will be explained in some more detail in Chapter 4, while the case for modular organisation is discussed widely in the IOMP Model Curriculum for postgraduate (MSc-level) education programme on Medical Physics [1], and in the experience of the author, it provides an optimal solution. While all this is useful to be taken into consideration, it is worth noting that the front page organisation can be modified according to individual circumstances; this being made possible by the fact that Moodle allows for a great flexibility in structure, appearance, and organisation of content. Examples of how this can be achieved will be discussed in Chapter 4. Again a reminder that it would be the Manager of the course, which would be either the Programme Director or a dedicated Course Administrator, but not the Course Creator at IT, who sets up the internal structure of the course and this structure should be in line with the Programme Specifications and the overall Regulations of the educational institution.

Getting back to the example shown in Figure 3.2, there are five sections in the main part in the centre of the front page: Summary, Lectures, Quizzes, Coursework, and Past Exam Papers (there may be a sixth one, Tutorials, or it can be incorporated in the fifth section to be renamed, e.g., Exam Preparations). This layout is suitable to present a condensed, uncluttered information at first glance for a complex subject, and in particular a subject that contains a lot of image-related information, typical for Medical Physics.

There may be additional blocks to the left and right of the centre to help with the navigation of the site and with the choice of settings. All of them can be minimised in a variety of ways when they are not needed. In some versions of Moodle some actions of minimising are referred to as 'decking' and would be useful in case the main structure feels very much overloaded with information. In principle, decked blocks would appear in the far-left upper corner. In later versions of Moodle, the function is simplified and minimisation is straightforward for instance when selecting/deselecting the 'down' arrow adjacent to each section in the left-side block in Figure 3.2 for its drop-down menu to be displayed or hidden, respectively.

It is helpful for the Summary to include an outline of the aim of the course, a list of recommended reading, and a method of assessment and formation of overall module (course) mark. Typically, the summary section can contain information about the aim (also known as 'learning objectives') of the course, the number of contact hours (or as a percentage of the total contact hours for the MSc programme), brief descriptions of the sections, and the method of assessment and formation of the grades. For instance, for the course of Imaging with Ionising Radiation (IIR) in our case, the quizzes section will provide assessment of the students' knowledge throughout the course; normally, it would appear immediately after the respective lecture is delivered; quizzes will be *optional*, and in case a mark is provided, it will be just to give an idea to the participants how they have mastered the material, i.e., the mark from the quizzes will not count towards the final mark for the module or to any other mark in the MSc programme. This is the reason why it is organised in a separate section and is not included in the Coursework one.

In contrast, the Coursework section provides an assessment which is *mandatory* to complete and will count towards a certain percentage of the final mark for the module. In our example, the weight of the coursework is 30%, which in the author's experience provides a balance between student involvement throughout the course and the weight of the unseen exam; however, this figure depends exclusively on the course regulations adopted at the educational institution.

In the example discussed, the Lecture section consists of two folders – Current lectures and Archive lectures – and there is a note reminding the Teacher that the Archive folder is invisible to the students. In our experience, the availability of this Archive is very useful and it serves to assist the Teacher and all role players with editorial functions – for instance before the start of the module the Manager can prompt the Teacher to upload their lecture/s for the coming session, sending them the archived lecture from the previous year as a reference which will most probably be in need of an update, as Medical Physics and Engineering is a very dynamic subject.

The module of IIR, used in the example, is delivered over an intensive period of 5 study days. This can be a typical organisation for a course in the specialist area of Medical Physics and Engineering especially concerning an MSc programme (and to a lower extent a BSc programme), dictated by the necessity of involving lecturers from clinical practice as well as other factors. A more detailed discussion about the necessity of intensively organised modular courses in Medical Physics is made in [1]. This article, containing also the IOMP Model Curriculum for postgraduate (MSc-level) education programme on Medical Physics, cited here several times, can be accessed for free at www.mpijournal.org/pdf/2013-01/MPI-2013-01-p015.pdf.

The sub-section 'Lectures IIR 2019-20' in the 'Lecture' section contains as an indicative example lecture handouts for this 5-day current period; they are available mainly as PDF files, with their titles reflecting the order in which a certain lecture is delivered (Figure 3.3). Usually the Manager will rename the PDF files with the handouts according to the schedule of delivery of the lectures. This will be described in Chapter 4.

The preparation of the lecture material for upload is important as it aims to minimise the space taken by the course without jeopardising the quality and pedagogical value of the lecture

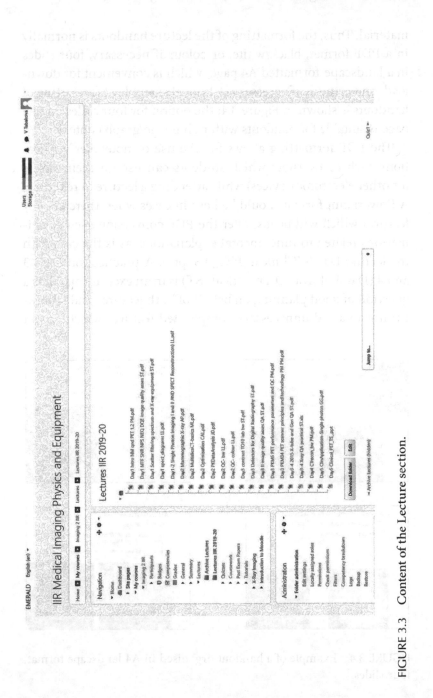

FIGURE 3.3   Content of the Lecture section.

material. Thus, the formatting of the lecture handouts is normally in a PDF format, black/white, or colour if necessary, four slides in a landscape-formatted A4 page, which is convenient for download, and printing if necessary. A typical example of formatting of handouts is shown in Figure 3.4; the option for four slides on one page is suitable for handouts with rich image/graph content.

The PDF formatting allows for the use of notetaker applications such as Evernote which students can use on their tablets (or other electronic devices) while attending a lecture in real time. A PowerPoint format should be kept in cases when there are key features which will be lost after the PDF conversion such as animations related to fundamental explanations, as is the case with the lecture Day 5 (Clinical_PET_TS_ppt). A practical on Days 3 and 4 (Day 3–4 Xray QA practical_ST) is in an excel format. It is a question of good planning on behalf of both lecturer and administrator to avoid unnecessary complicated features which do not

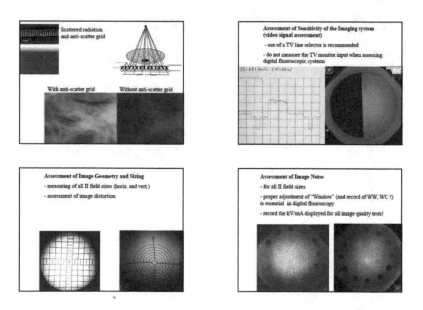

FIGURE 3.4   Example of a handout organised in A4 landscape format, four slides.

contribute to the pedagogical value of the lecture. Another consideration that has to be kept in mind is the Copyrighted material included in the lecture notes. In the light of its importance and its additional dimension in the internet application, the steps necessary to undertake in order to conform to copyright requirements are discussed in detail in Chapter 2 and here is a quick check-in list:

- First check if the copyrighted material is covered by a university-broad licence, such as one issued by the CLA (Copyright Licensing Agency) in the UK or similar.

- If this is not the case, seek permission from the copyright owner (which can be the author, their publisher, or a third party). In case of doubt contact the author directly and they will be able to help locate the copyright owner. Another source of such information is the legal department of the University/company where the author(s) is/are employed.

- In any correspondence with the copyright owner, highlight the fact that the material will be available for limited access only, to a limited number of users on a secure platform (and not freely with unlimited access on the internet).

- In any case after receiving the copyright permission, do not forget to acknowledge the copyright (i.e., the figure/diagram is reproduced by permission of…/courtesy of…).

For detailed steps on the preparation of the content, see also Section 2.4 – The Moodle VLE – Teacher Preparation.

3.4 THE EDITING FUNCTION: A MAIN TEACHER FUNCTION

The Teacher is authorised to add material and, in general, to edit the content of the course. This is achieved by the Editing Function. It can be accessed by selecting the 'Turn editing on' option which

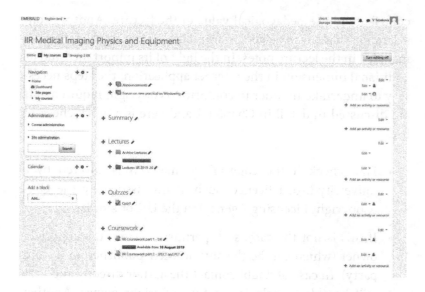

FIGURE 3.5   Front Page with the editing function enabled.

appears at the top right corner of Figure 3.2. It should be noted that Moodle can be customised in a slightly different way and then this function can appear in a Settings block, if available separately, or as a cogwheel, usually at the top left corner. In any case the access will be easy and intuitive. No matter how and where the function is selected, the result will be identical to the one shown in Figure 3.5.

As can be seen in Figure 3.5 the editing function appears on several levels. Thus, it is possible to manage a whole section (topic) or part of it.

3.4.1 Editing a Section

We shall take as an example the managing of the section 'Lectures'. In Figure 3.6, using the button 'Edit' to the far right of the section 'Lectures' the variables that can be manipulated are as follows: the topic 'Lectures' can be edited, highlighted, hidden from student view (in which case the symbol of visibility, the eye, will appear with a cross through it), or deleted (the sign of the 'bin').

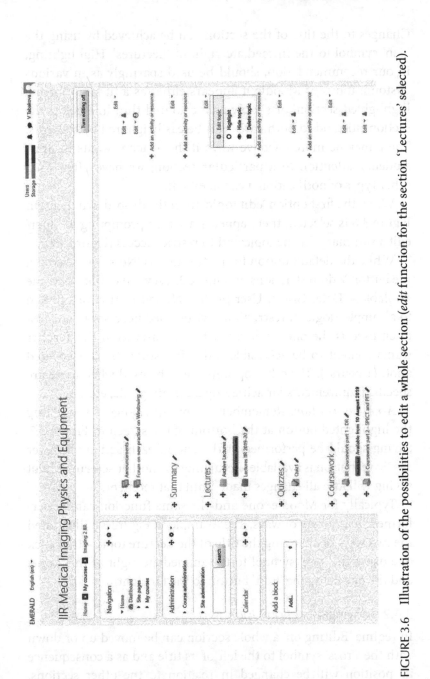

FIGURE 3.6 Illustration of the possibilities to edit a whole section (*edit* function for the section 'Lectures' selected).

Changes to the title of the section can be achieved by using the 'pen' symbol to the immediate right of 'Lectures'. Highlighting, in our recommendation, should be used sparingly as in various customisations of Moodle it appears either as a lit lamp for the highlighted section, or simply as a coloured thin line below the section title and in both cases can be easily ignored by students. There may be more effective ways if the Teacher wants to draw students' attention to a particular section; we have often used other types of notification, i.e., via e-mail.

When the first option 'edit topic' from the drop-down menu in Figure 3.6 is selected, there appears a screen prompting to enter/edit a summary for the topic and to restrict access (Figure 3.7).

While the default option for restrictions is 'None', by selecting the button 'Add restrictions' in Figure 3.7, several options become available – Date, Grade, User profile, Nested restrictions linked to a complex logic. If restrictions are deemed necessary, the most often used is the one by date (useful for instance if the Teacher wants content to be accessible only after students have covered another course). The other options are to be used when there are specific requirements for achieving a certain grade, etc.

A note of caution: Remember to save all changes by selecting the highlighted button at the bottom of the screen in Figure 3.7. Saving should be performed at the end of each action wherever the 'Save' button is available. Returning to another screen without saving will lose all changes made until that moment.

Typically for Moodle, one and the same function can be performed in a variety of ways. The example in Figure 3.6 illustrated the two ways of changing the title of the Lecture topic – the quick one using the 'pen' symbol to the immediate right of the name; and the one using the 'Edit' button to the far right.

3.4.2 Moving a Section Up/Down

In regime 'Editing on' a whole section can be moved up or down with the 'cross' symbol to the left of its title and as a consequence its position will be changed in relation to the other sections.

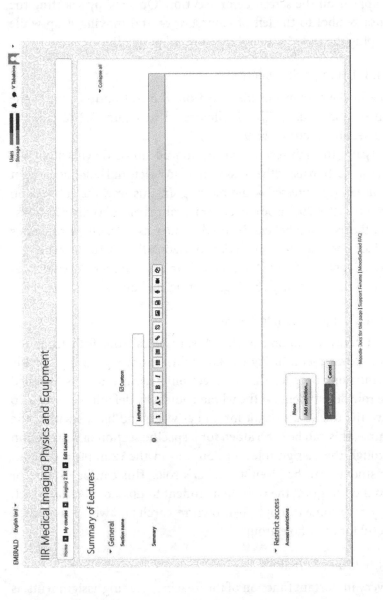

FIGURE 3.7 Managing the Lecture topic.

(i.e., in Figure 3.6, the 'Coursework' section can be moved up to appear on the screen before section 'Quizzes' by selecting the cross symbol to the left of Coursework and moving it up while keeping the right button of the 'mouse' pressed).

3.4.3 Editing a Sub-Section

To edit the contents inside a section, the edit button next to the sub-section is used. This is illustrated in Figure 3.8 to edit the sub-section Lectures 2019-20.

Again, the sub-section can be moved up or down within the Lecture section using the cross symbol to its left and holding the right mouse button pressed while moving. In this way, the sub-section Lectures 2019-20 can be moved for instance before 'Archive lectures', and the 'pen' symbol can be used to alter its title. Other functions include moving the sub-section horizontally (useful to achieve a visual representation of hierarchical structure if more levels of sub-sections are necessary), hiding, deleting, and duplicating.

3.4.4 Assigning Limited Roles

An interesting option is the 'Assign roles' (the fifth from the drop-down menu in Figure 3.8), which allows granting specific permissions only for this sub-section – i.e. if a user is assigned the role of a student for the whole course, by default they will also have the role of a student for all activities and blocks within the course; this can be overridden for a specific section or sub-section through the 'Assign roles' option and in the example a particular student can be given a teacher's role. This can be useful, for instance, to grant the right to a student to upload material which they have come through their own research and which is deemed useful for the whole group.

3.4.5 Other Functions

A very important function of the Teacher, grading assignments, is discussed in Section 4.3 after the illustrations of Students submitting assignments.

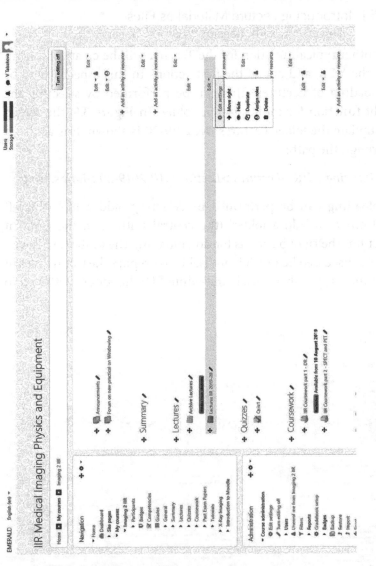

FIGURE 3.8   Editing the content for the sub-section Lectures 2019-20.

3.5 CREATING CONTENT: ANOTHER MAIN TEACHER FUNCTION

3.5.1 Introducing Lecture Material as Files/ Folders and Other Resources

Another typical function for the Teacher will be creating content on the VLE, and one of the first actions in this aspect would be uploading of Lectures. This can be performed by selecting the Edit function for the respective folder. In Figure 3.9, the screen to update the folder Lectures IIR 2019-20 is shown. It is accessed through the path:

Imaging 2 IIR >Lectures >Lectures IIR 2019-20 >Edit settings

Uploading can be performed by selecting 'add a file' (the left button) or a 'add a folder' (the central button) in the Content section (the right button is for downloading the available files).

As discussed before, it is advisable to prepare the lecture files in advance, typically as black-and-white PDF handouts with two to

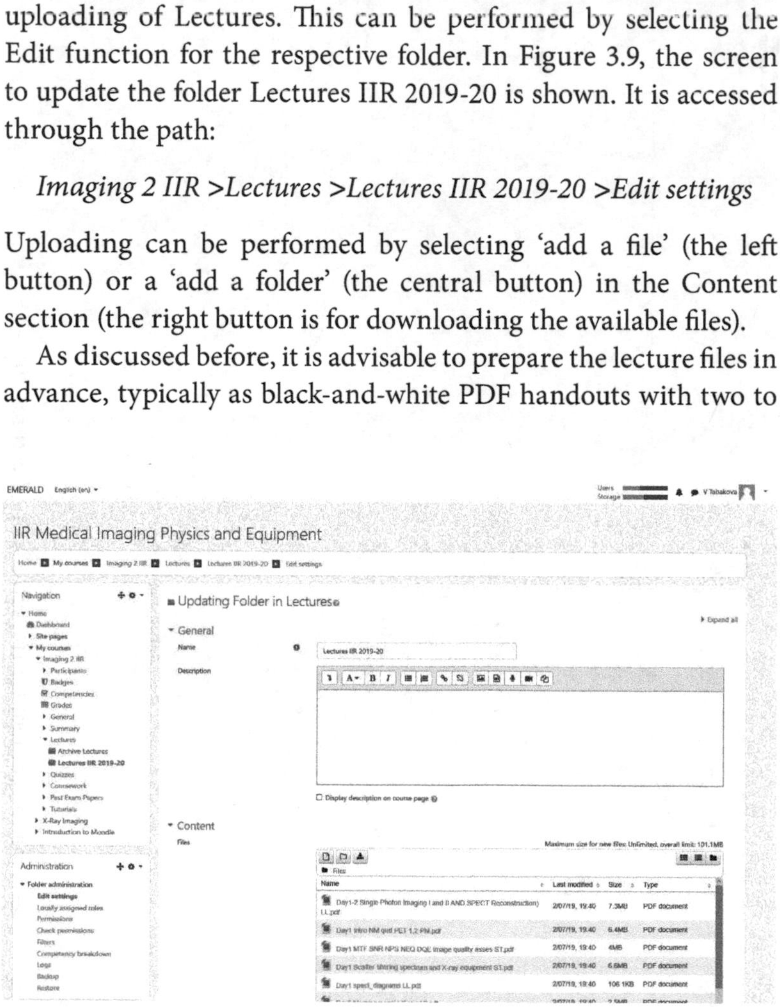

FIGURE 3.9   Screen for uploading lecture files.

four slides per page, their name reflecting the order in which the lecture appears in the syllabus and to store them in the computer used to manage the VLE.

Upon selecting the 'add file' button (first left in the Content section in the centre in Figure 3.9), the 'File picker' option appears (Figure 3.10).

The lecture file will be uploaded as an attachment by selecting the 'Browse' button on Figure 3.10.

If for any reason (i.e., updating) it is necessary to delete a lecture, this can be performed by 'right click' on the name of a lecture on Figure 3.9. The lectures are arranged in alphabetical order automatically, so it is advisable to name them in such a way so that this is taken into account and corresponds to the schedule of delivery – in the example all names start with 'day' and the number of the day on which the schedule is delivered. In reality the name of the first file can begin with the first date of delivering the modular course – i.e., 15.1.20, then the

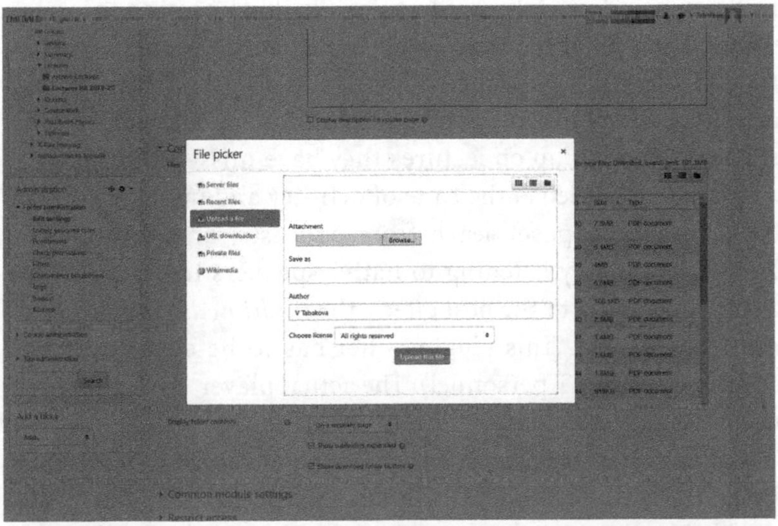

FIGURE 3.10   File picker.

abbreviated name of the lecture, followed by the initials of the lecturer. When other lectures follow the same pattern of naming, the lectures will be arranged according to the schedule of delivery. In case there are any changes to the schedule, changing the date at the front of the name of the lectures concerned will lead to their automatic re-arrangement. This pattern obviously is convenient for a course where the lectures are delivered in a modular way.

Files and folders can also be introduced by selecting the 'Add resource/activity' button at the bottom of section Lectures on the front page with function 'Editing' on (i.e., in Figures 3.5 and 3.6). The result is a drop-down menu that lists a variety of resources which can be introduced – files, folders, URL. For instance, if the Teacher needs to provide a web link as a course resource, its URL can be added by the method described above. The URL will appear on the front page in the corresponding section. A major consideration here is securing copyright and the steps discussed previously in Chapters 2 and 3 should be followed. The link can be embedded or alternatively could open in a new window. Advanced options include passing information to the URL if required (such as a student's name). Another type of files which can be added is audio (sound) files. Using audio files is a very powerful tool in a Moodle course, allowing students to catch up on lectures they have missed (the lectures need to be recorded at the time of delivery and uploaded on the VLE for the purpose); learn from podcasts, or improve their language skills by listening to native speakers interact. To use audio in Moodle to the best effect, the *multimedia plugins filter* must be enabled. This plugins filter has to be enabled by site administrators (IT personnel). The actual player resides on each user's computer.

For example, a teacher may put a MP3 audio file as a resource in their course, or have a URL link to an external MP3 file. When the MP3 audio plugin has been turned on, the student will be able to play them in Moodle using a media player on their computer.

For playback of these audio files the students (users) should have Adobe Flash, QuickTime, Windows Media Player, and Real Player installed on their computer. If users do not have these installed they may be prompted to go and install them by their browser. These pieces of software are generally free, easily installed, and widely used.

Similarly video files may be embedded with the *Multimedia plugins* filter enabled.

WORD OF CAUTION: When uploading both audio and video files, attention has to be paid about the size of the file; depending on the package of Moodle (discussed in Chapter 5), there may be restrictions to the overall size; in any case, it is good housekeeping to delete big files when they are made redundant (for instance when an audio file of a pre-exam tutorial is uploaded, it should be discarded after the exam).

3.5.2 Creating Coursework Assessment

The Teacher often will deal also with setting up a new Coursework. This is performed on the main page of the course with turned Editing on (Figures 3.5 and 3.6). A very useful function is the 'add an activity or resource' one, to be found at the right bottom corner of each section. This activity, apart from providing another way to add files and folders to the lecture section, is also the way to organise a coursework (selecting the first option from the drop-down menu shown in Figure 3.11).

The next step is selecting the Assignment option (highlighted in Figure 3.11).

The setting up of an assignment requires a name, description of the task (in the section General, identical to the two sections of a quiz as shown later in Figure 3.13), and defining the availability of the assignment – for instance the assignment can be made available only after a certain block of lectures has been delivered (Figure 3.12).

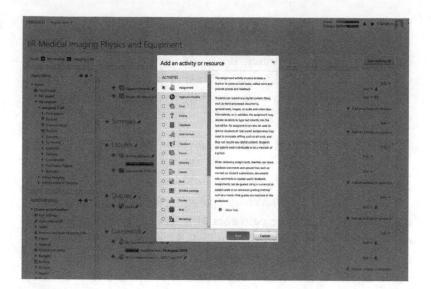

FIGURE 3.11   Adding a coursework.

FIGURE 3.12   A screen for adding a new assignment to the coursework section.

If the cut-off date and time is set, submissions of coursework will not normally be accepted by the site after this date and time, unless an extension is given. The procedure for giving an extension would be defined in the Programme Regulations of the educational institution, but is in principle started individually for a student who has requested an extension within a defined period of time and who has submitted the necessary documentation in support of their request. The form for submitting such a request is often referred to as 'Mitigating Circumstances Form'. After the extension is approved by the authorised person or committee, the Manager will set the new date of submission individually for the student concerned. The steps for setting up this new submission closing date are discussed in Section 4.1.12.

3.5.3 Other Forms of Assessment

Quizzes are another very useful form of assessment, and they allow the Teacher to include in them questions of various types (i.e., multiple choice, numerical, etc.). They provide a platform for ongoing assessment of the extent to which the student has mastered the material taught so far. The teacher can allow the quiz to be attempted multiple times, with the questions shuffled or randomly selected from the question bank. A time limit may be set. Each attempt is marked automatically, with the exception of essay questions, and the overall grade is available to view in the Grader Report, which for the course IIR can be accessed via *Imaging 2 IIR > Grades > Grade administration > Grader report* (the link provides access to all grades including the ones for coursework and the Teacher can view grades for all students).

For the quizzes, the Teacher can choose when and if hints, feedback, and correct answers are shown to students.

In principle, quizzes may be used

- As mini tests for reading assignments or at the end of a topic.

- As exam practice using questions from past exams.

- To deliver immediate feedback about performance.

- For self-assessment.

- As course exams (rarely).

In our case in the course of IIR the Quizzes section will provide assessment of the students' knowledge throughout the course; the quizzes will become available after the respective lecture is delivered; quizzes will be **optional** and in case a mark is provided it will be just to give an idea to the participants how they have mastered the material; i.e., the mark from the quizzes will not count towards the final mark for the module or towards any other mark in the MSc programme. This is the reason why it is separated from the next section – the Coursework.

To set up a quiz, the button 'Add an activity/resource' is selected from the front page (see Figures 3.5 and 3.6) and the screen of Figure 3.11 appears. By selecting Quiz from the drop-down menu, there appears the screen illustrated in Figure 3.13. The first step of setting up of a quiz is very similar to creating a Coursework and the section 'General' in both cases contains the blocks *Name* and *Description*. There follow the sections *Timing* and *Grade*. The method of Grading would differ from a coursework mainly when multiple attempts in a quiz are allowed. Then the methods for calculating the final quiz grade can vary from highest grade of all attempts through to average (mean) grade of all attempts; first or last attempt (all other attempts being ignored).

Timing refers to the dates of opening and closing PLUS time limit to perform the quiz – these dates can be set manually and in conjunction with the delivery of the relevant lectures, which provide material for the quiz. The time limits to perform the quiz can be set in minutes (default setting), seconds, hours, days, and weeks.

Further down, the Teacher/Manager is prompted to select the number of attempts allowed, the layout (mainly to determine the question order, how many questions would appear on every page,

FIGURE 3.13   Constructing a new quiz – description and timing.

and whether students are restricted to progress through the quiz in order or are able to return to previous pages and/or skip ahead), the mode of feedback, etc. An important step is setting the 'Review options' (which can be during the attempt, immediately after the attempt, or when the attempt is finished), reached by scrolling down and visible on Figure 3.14 (which is a continuation of the screen shown in Figure 3.13 and is reached by scrolling down).

The review options control what information students can see when they review a quiz attempt or look at the quiz reports and they are:

- 'During the attempt' settings are only relevant for some types of questions (i.e., interactive with multiple attempts allowed).

- 'Immediately after the attempt' settings apply for a set number of minutes after the submit button is clicked (this is pre-set at 2 minutes by the Moodle programmers, so there is no need to do an adjustment).

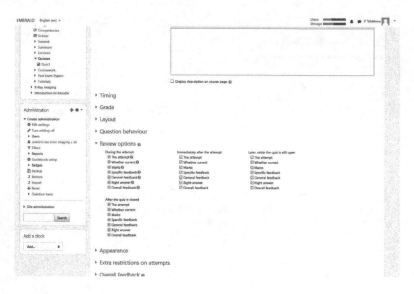

FIGURE 3.14   Constructing a new quiz – review options.

- 'Later, while the quiz is still open' settings apply after submission, but before the quiz' closing date; the last option is to look at the quiz reports only 'after the quiz is closed'.

The section marked as 'Question behaviour' is collapsed on Figure 3.14 and, when selected, gives the following setting options (the first two of which are visible on Figure 3.15):

- *Shuffle within questions*: This setting applies to questions that have multiple parts, such as multiple choice or matching questions, and the option must also be enabled in the individual question's settings. By default, this is set to **No**. If set to **Yes**, each time a student attempts the quiz, the parts of multiple choice (or other multiple part questions such as matching questions) will be randomly shuffled. For example, in a multiple-choice question, this setting will randomly shuffle the answers for choices A, B, C, D, etc., so that

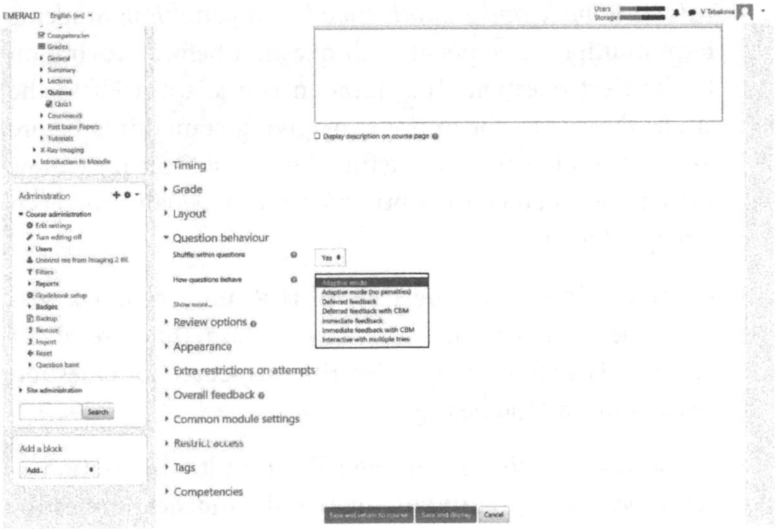

FIGURE 3.15   Constructing a new quiz – question behaviour.

the correct answer will be in a different location for each attempt. This option is an important feature and the set up again depends on the pedagogical aim.

- *How questions behave*: These settings control when and how students receive feedback.

- *Deferred feedback (Default)*: Students must submit the quiz before grading and before receiving feedback.

- *Interactive with multiple tries*: Students receive immediate feedback as they submit each question; if they do not answer correctly, they can try again; there is an additional option for penalty through losing points for incorrect tries, which can be done in the individual question settings.

- *Immediate feedback*: The student can submit their response to a question immediately and get it graded, but can submit only once and cannot change the response.

- *Adaptive mode and Adaptive mode (no penalties)*: Students have multiple attempts at each question before moving on to the next question. The question can adapt itself to the student's answer, for example by giving some hints before asking the student to try again. The two sub-settings allow for choice whether to subtract/or not a penalty for each wrong attempt.

- *Certainty-Based Marking* (CBM), is a more complex system where students indicate how certain they are about answers. For more detailed descriptions search for Question Behaviour on Moodle.org.

- *Allow redo within an attempt*: By default, this option is set to **NO**. If set to **Students may redo another version…**, when students have finished attempting particular question, they will see a Redo question button. This allows them to attempt another version of the same question, without having to submit the entire quiz attempt and start another one. This setting only affects questions (for example not Essay questions) and behaviours (for example immediate feedback or interactive with multiple tries) where it is possible for a student to finish the question before the attempt is submitted.

- *Each attempt builds on the last*: By default, this setting is disabled. If multiple attempts are allowed (earlier in the settings under **Grade**) and this setting is enabled, each new quiz attempt will contain the results of the previous attempt. This allows a Quiz to be completed over several attempts.

Another important feature is organising feedback in quizzes. This is reached by selecting 'Overall Feedback' in Figure 3.15 and the result is shown in Figure 3.16.

The overall feedback is the text which appears after a quiz has been attempted. In principle by specifying additional grade

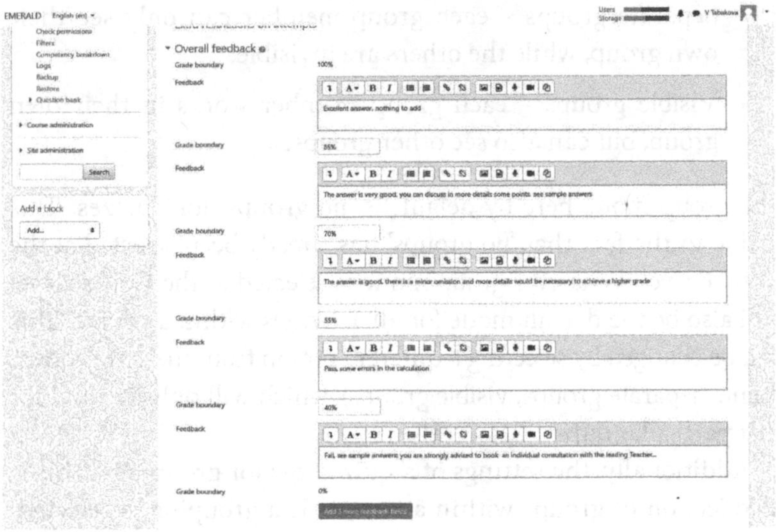

FIGURE 3.16 Constructing a new quiz – setting grade boundaries and feedback.

boundaries (usually as a percentage or as a number), the text shown to the student can depend on the grade obtained. An example of feedback for a quiz graded in percentage with a 15% step would read as follows: 100% – excellent answer, nothing to add; 85% – the answer is very good, you can discuss in more details points a and b; 70% – the answer is good, there is a minor omission in point c and more details would be necessary in points f and g for a higher grade; 55% – pass, section d contains an error in the calculation of…; the discussion would benefit from…; see also sample answers; 40% and below – fail (see sample answers; you are also strongly advised to attend consultation … on date).

The 'Common module settings' from Figure 3.15 when selected allows for creating groups (Figure 3.17). When this function is selected, three options appear:

- No groups – there are no sub-groups, and everyone is part of one big group.

- Separate groups – each group member can only see their own group, while the others are invisible.

- Visible groups – each group member works in their own group, but can also see other groups.

The group mode here by default is 'no groups' for quizzes. This is due to the fact that 'no groups' has already been selected at the Course level. Whatever group mode is selected at the Course level will also be the default mode for all activities within a course. This can be changed by selecting a different option from the scroll-down menu (separate groups, visible groups), which will only be valid for that particular quiz.

Additionally, the settings of a quiz allow for grouping – this is a collection of groups within a course. If a grouping is selected, students assigned to groups within the grouping will be able to work together.

The Checkbox for availability for group members only is used to allow (if selected) the relevant activity or resource to become

FIGURE 3.17 Constructing a new quiz – common module settings.

available only to students assigned to groups within the selected grouping.

Also shown in Figure 3.17, the possibility to restrict access to the quiz is displayed; the default option being 'none'. The other options include 'Access from/to dates' determining when students can access the activity via a link on the course page.

The difference between access from/to dates (reached by selecting 'Restrict access' shown in Figure 3.17) and availability settings for the activity (achieved by selecting the dates in 'Timing' shown in Figure 3.13) is that outside the set dates in the 'Timing' section students will be allowed to view the activity description, whereas access from/to dates prevents access completely if outside the defined period.

In Figure 3.17, it is also possible to choose a restriction for Grade. (This is reached again via the drop-down menu when 'Restrict access' is selected.) Introducing such a restriction will determine any grade condition which must be met in order to access the activity – that means that the student is required to achieve a specific grade before accessing the quiz.

Multiple grade conditions may be set if necessary. In such a case the activity will only allow access when ALL grade conditions are met.

The final step as usual includes saving all changes in the system. This is achieved by selecting one of the two highlighted buttons – 'Save and return to the course' or 'Save and display' in Figure 3.17. Returning to another screen without saving will lose all changes made until that moment for the quiz.

In brief – quizzes are a very flexible kind of assessment and the individual choice of setting will depend ultimately on the pedagogical paradigm.

3.5.4 Forums and Chats

The forum activity enables participants to have asynchronous discussions i.e., discussions that take place over an extended period of time. This activity is reached through the menu shown in Figure 3.11, when selecting 'Forum'. Then the following screen appears (Figure 3.18).

The only mandatory field is the title of the Forum. There are several forum types to choose from, such as a standard forum where anyone can start a new discussion at any time; a forum where each student can post exactly one discussion; or a question and answer forum where students must first post before being able to view other students' posts. The Teacher can allow files to be attached to forum posts. Attached images are displayed within the forum post. The Forum can be made available only to selected participants through Common Module settings (i.e., hidden from students, available only to other Teachers). Restricting access can be introduced by date or entry requirements (achieving a certain grade before accessing the forum). These actions are similar to the ones illustrated for quizzes in the previous Section 3.5.3. It should be noted here that starting a Forum is a function allowed to the Teacher (and also the Manager); students cannot start a new Forum, they can only participate in existing ones.

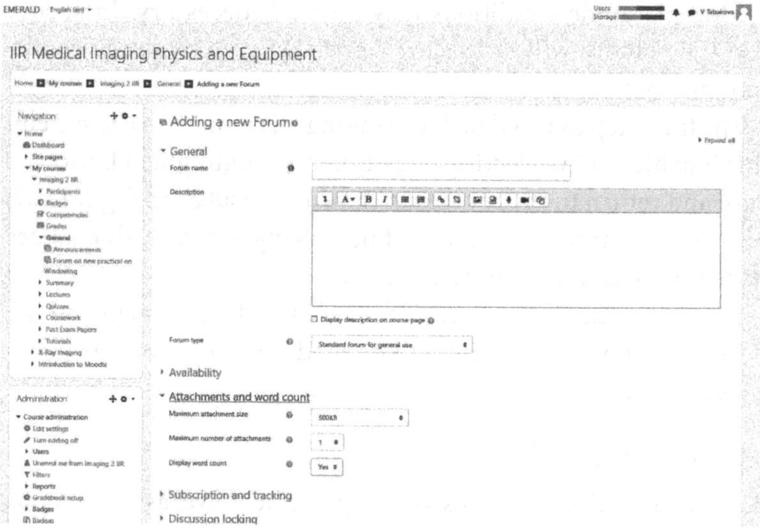

FIGURE 3.18   Organising a forum.

Participants can subscribe to a forum to receive notifications of new forum posts. A Teacher/Manager can set the subscription mode to optional, forced or auto, or prevent subscription completely. If required, students can be blocked from posting more than a given number of posts in a given time period; this can prevent individuals from dominating discussions. As before, to introduce the desired changes, they need to be saved at the bottom of the scroll page.

The Chat activity differs from the Forum one as it enables participants to have text-based, real-time synchronous discussions. This activity can again be reached through the drop-down menu shown in Figure 3.11 and is managed through the screen shown in Figure 3.19. Similar to the Forum, starting a Chat is a function allowed to the Teacher (and also the Manager); students cannot start a new Chat, they can only participate in existing Chats.

Again, the only mandatory field is the name of the Chat.

The Teacher can set up the chat as a one-time activity or as a repeated one at the same time each day or each week through

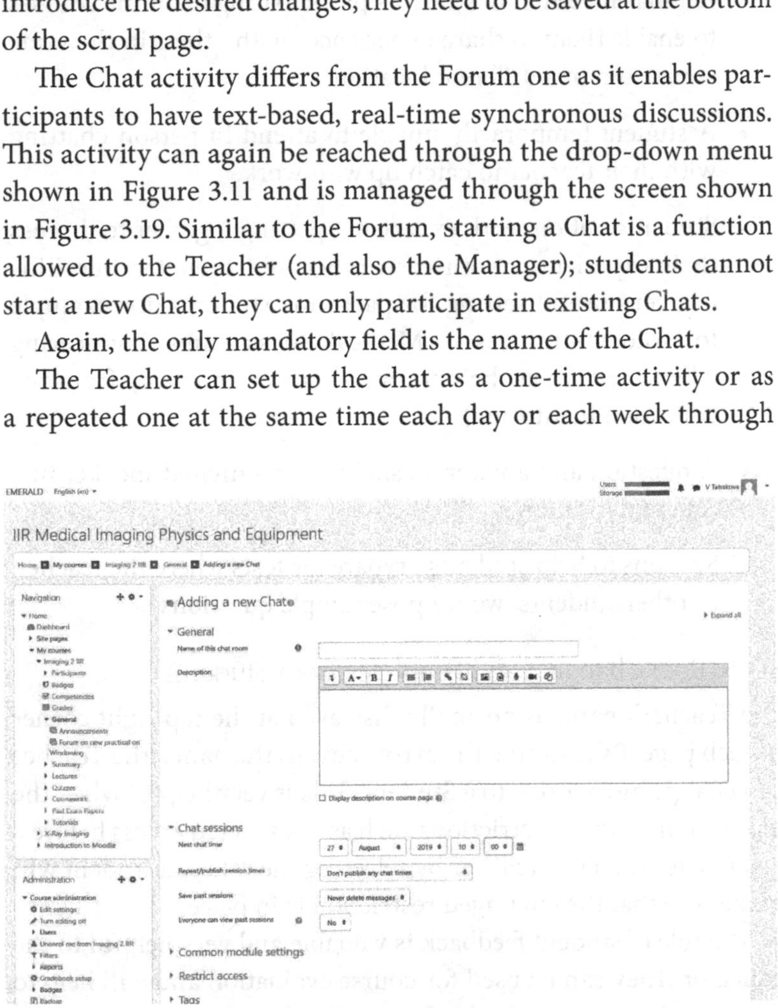

FIGURE 3.19 Organising a chat.

'Chat sessions'. The chat sessions are saved and can be made either available for everyone to view or restricted to users with the capability to view chat session logs.

Chats are especially useful when the student group is not able to meet face-to-face, such as

- Regular meetings of students participating in online courses to enable them to share experiences with others in the same course but in a different location.

- A student temporarily unable to attend in person chatting with their teacher to catch up with work.

- Students out on work experience getting together to discuss their experiences with each other and their teacher. This is a very useful setting for part-time students (for example for student–trainees in Medical Physics and Engineering when they are at their training centres and away from the education institution).

- A question and answer session with an invited speaker in a different location.

- Sessions to help students prepare for tests where the teacher, or other students, would pose sample questions.

3.5.5 Role Changing and Feedback from Students

The Teacher's name is normally displayed at the top right corner of each page. By selecting the arrow next to the name, the Teacher can change his/her role to a Student. This is very helpful when the goal is to introduce restrictions such as a restricted access before a given date. Then the Teacher by selecting the role of a student will make sure that the envisaged restriction is in place.

The role of student feedback is valuable and very helpful to the educator. They can be used for course evaluation and will help to improve the content for the future. Furthermore, there may be institutional requirements for collecting feedback from students.

This is an activity which can be organised by the Teacher. The steps to follow are as follows: Select 'Add activity/resource' at the bottom of any section from the front page with 'Editing' turned ON (Figures 3.5 and 3.6). The screen that will be reached is the one shown in Figure 3.11. From this screen 'Feedback' is selected. The Teacher can thus create a custom survey for collecting feedback from students. The question types can be multiple choice, yes/no, or text input.

BIBLIOGRAPHY

1. Tabakov, S., Sprawls, P., Krisanachinda, A., Podgorsak, E., and Lewis, C. IOMP model curriculum for post-graduate (MSc-level) education programme on Medical Physics. *Journal Medical Physics International* 2013; 1: 16–22, available from: www.mpijournal.org/pdf/2013-01/MPI-2013-01-p015.pdf.
2. IAEA Training Course Series No. 56, Postgraduate Medical Physics Academic Programmes, 2013, available from: www.iaea.org/publications/10591/postgraduate-medical-physics-academic-programmes, last accessed 1 Dec 2019.

CHAPTER **4**

Role-Specific Functions on Moodle

4.1 THE ROLE OF THE MANAGER: CREATING A COURSE STEP BY STEP

The typical role of the Manager is to organise the programme as a whole as it appears on the Virtual Learning Environment (VLE), in other words to organise and manage the structure of a programme [which in our example will correspond to the curriculum of an MSc Programme in Medical Engineering and Physics (MEP)]; and to this effect, he/she will work in close collaboration with the Programme Director.

Additionally, the Manager will enjoy the same content building rights as Teachers, as described in detail in Chapter 3. However, his/her role in creating lecture and coursework content will normally be minimal, with the exception of managing handouts.

4.1.1 Managing Handouts

The lectures from various Teachers will either be uploaded by them directly on the VLE platform or will be sent to the Manager. The latter option will have the following advantages: the Manager will see to it that PowerPoint presentations will be uploaded only when it is absolutely necessary (for example if there are important features and information which will be lost when converting a PowerPoint file to a PDF one). This is not only a sign of good housekeeping, but will also help keep down the size of uploads to the VLE platform, especially important if there is overall data volume restriction (for instance as discussed in Chapter 5, esp. with regard to Table 5.2). The other advantage is that the Manager will typically rename the PDF files with the handouts according to the schedule of delivery of the lectures.

A recommendation for the titles of the lectures is that they should be made uniform starting with the date of delivery, followed by an abbreviation for the lecture title and containing at the end the initials of the lecturers. In reality, the titles received from the lecturers will vary, and it is recommended that the Manager make the titles uniform, as this will be useful for future reference.

Then the Manager would upload the uniform lectures. For guidance on uploading lectures and other files, the explanations in Chapter 3 will be very useful.

4.1.2 Setting Up a Profile

In most cases once the Manager enters their role, s/he becomes the face of the Programme and often the first point of contact for all other users (students, teachers, guests, etc.) As such, it is very important that the Manager sets up a profile on the VLE with as much information as relevant. Setting up a profile is performed after logging (for reference on logging-in see Figure 3.1 and following explanations) the sequence Dashboard > Preferences is followed or via selecting directly on the Name of the user > User account > Profile. (The name of the user will be at a prominent place on all screens after logging, and in the set-up of the examples in this book is always at the right top corner.)

The required fields when setting up a profile are as follows: username, first name, surname, and e-mail address. There is an option to make the user's e-mail visible to all. (Note that privileged users such as Managers or Teachers will always be able to view the e-mail of any user.) It is worth mentioning here that it is better to upload a user picture when setting up a profile; for the current version of Moodle (2019) if a profile picture is not uploaded, Moodle will attempt to load a profile picture from the web service 'Gravatar'. (The acronym "Gravatar" stands for Globally Recognised 'Avatar'. In the computer world, 'Avatar' denotes a character that represents the online user and 'Gravatar' is a web service that allows users to upload such a character online and become part of an online community; the 'avatar' becomes associated with the online user's e-mail address.) In short, at the time of writing if a user has not created an online 'Avatar' on the Gravatar website, and does not upload a picture on Moodle, no profile picture will appear in the user's profile.

At the end it should not be forgotten to save the profile changes with the dedicated button which will appear at the bottom of the screen (this button is not visible in Figure 4.1, but is similar to the 'save' buttons on Figures 3.7, 3.17, and elsewhere).

The same sequence for setting a profile is valid for all users – they all, without exception, have the right (and should be prompted by the Manager at the start of the course) to set up

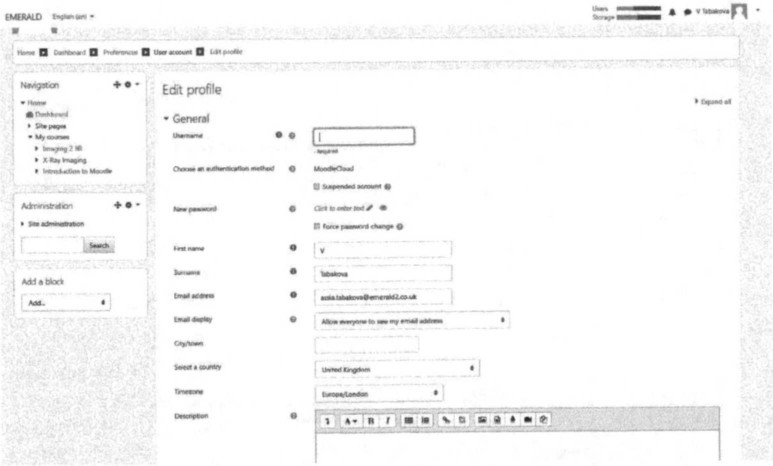

FIGURE 4.1   A sequence of setting up a profile.

profiles; this is especially useful for large courses and for courses including a lot of external tutors.

4.1.3 Creating the Structure of the Programme

A very important function of the Manager is creating the structure of the Programme (in our case this should correspond to the curriculum of the MSc in Medical Engineering and Physics). A front page of the whole programme would include all courses on this curriculum; a selection of these courses is included in Figure 4.2; typically, a similar screen will appear after log-in (Figure 3.1) and will contain all courses a user is enrolled to. Note that depending on the device for viewing and on the number of items it may be necessary to scroll down to view all items on the front page. In the example all available courses are listed on the figure. (This is a selection and NOT the complete list of courses in an MSc programme in Medical Physics.)

4.1.4 Adding a New Course

A button 'Add a new course' is available at the bottom of Figure 4.2 when the editing function is selected (for reference on the Editing function, see Figure 3.5 and following explanations in Chapter 3).

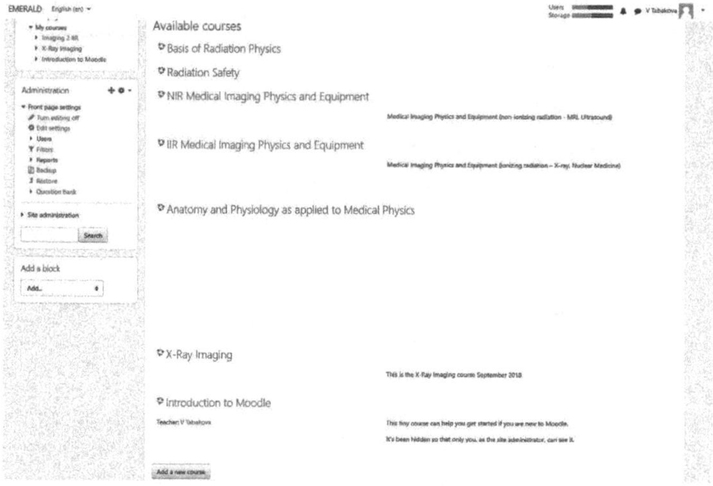

FIGURE 4.2   EMERALD Medical Physics – available courses – selection.

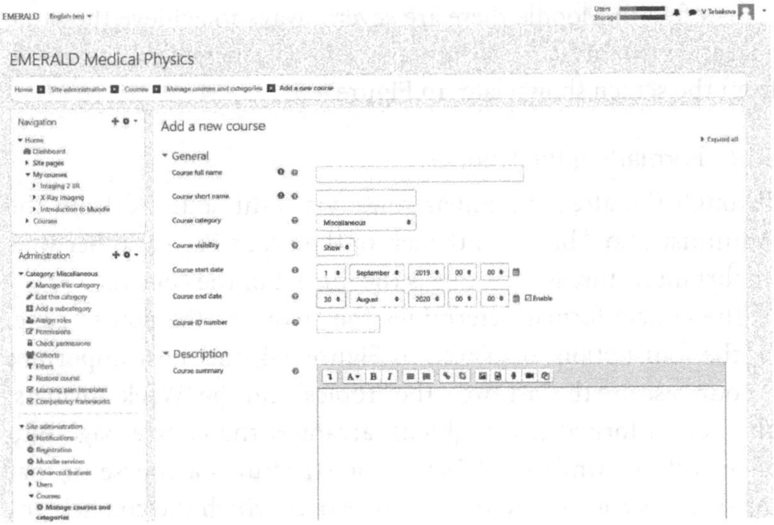

FIGURE 4.3   Adding a new course.

This button enables the Manager to add more courses and, when selected, the screen in Figure 4.3 appears.

The procedure for adding a new course requires entering the Course full name and a Course short name, – the first two fields in Figure 4.3.

These two fields are the only mandatory ones; the full name should correspond to the approved curriculum for the Programme and the short one should be agreed with the Programme Director.

4.1.5 Managing Visibility

Once the courses are all introduced (or at any point before that) they can be manipulated/managed – i.e., made visible/invisible for other users (students, guest visitors, etc.,) which can be achieved with the function 'Editing turned on' when at the **Course Front Page** (i.e., IIR Medical Imaging Physics, Figure 3.5, Chapter 3), then selecting 'Edit Settings' from the Administration – Course Administration drop-down menu, and selecting the preferred option within Course Visibility (the default option being 'Show').

As often in Moodle there are several ways to achieve the same goal and visibility can also be managed very efficiently and quickly from the screen shown later in Figure 4.5.

4.1.6 Formatting the Course

Through the already familiar route via 'Edit settings' from the 'Administration' block on the left of the Front Page another very important feature is set – this is the format of the course.

The course format determines the layout of the course page. Of the four options displayed in Figure 4.4, the most important for our case are the last two - the 'Topics' and the 'Week' formats. The Topics format (highlighted), arranges the course page into topic sections while the Weekly one structures a course so that the sections correspond to the weeks over which the course runs and typically the first week should start on the course start date. The Topics format is suitable for courses in a modular form, with a majority of visiting lecturers, when the module is delivered in a limited period of time and not over the whole term, while the

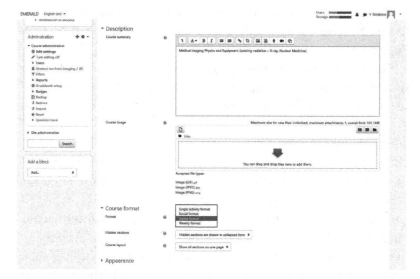

FIGURE 4.4   Choosing course format.

Weekly format is most suited to courses delivered over the length of a whole term. As such, the Topics format is the most relevant for Master courses with plenty of visiting lecturers from clinical practice, such as the MSc in MEP, as well as for any short-term modules such as courses for continuing professional development (CPD). The Weekly format is typically useful for BSc courses. For further discussion about the modular form of MSc courses in Medical Engineering and Physics see Chapter 3. We are using the Topics format in the examples in this book; however, course management will be similar in a Weekly format.

4.1.7 Managing the Order of Appearance of Courses

Apart from adding new courses and managing their visibility, sometimes it is important to bear in mind the order of appearance of courses. This is managed through the 'Course and category management' screen (Figure 4.5). The screen is reached by selecting Site Administration – Courses – Manage courses and categories.

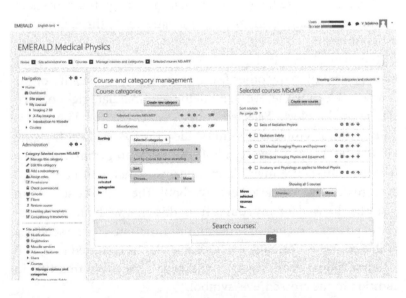

FIGURE 4.5 Managing the order of appearance of courses.

The move up/move down function (managed through the up and down arrows within the right block 'Selected courses MScMEP' shown in Figure 4.5 is often a good way to draw users' attention to a specific course. In our experience this proved particularly useful when the trainee students at an academic institution were in their final part of academic study and, having covered all subjects on their curriculum, had access to a number of courses (in our case, 12 separate courses), which would appear on the first page when they logged in (and depending on the size of the screen of the device they were accessing it through, view of all courses would involve several scrolls).

However, in this final stage the student-trainees were expected academically to participate in one single modular course, the last one on their curriculum – the MSc Project. Parallel with the preparations of their MSc thesis they were working as full-time trainees in their hospitals in various cities. The Manager of the VLE could help their busy schedule by securing easy access to the MSc Project course, by moving it to the top of the list of courses. A Manager can use his/her discretion to arrange the courses in a suitable way and to monitor this arrangement throughout the programme.

The view shown in Figure 4.5 is a very useful way of quick management of many other features, such as the visibility of the course – selecting the 'eye' symbol to the immediate right of each course in the block 'Selected courses MScMEP' on Figure 4.5. Thus, the only visible course for students is IIR Medical Imaging Physics [with open 'eye' symbol; the remaining four from the selected courses are invisible to students (which will be marked by a crossed 'eye' or 'closed eye' symbol)]. The visibility status can easily be changed for any course from this view without the necessity to go inside each course to edit this function. On the screen, a visible course is normally recognisable via the coloured title, and a course hidden from students will have a grey-coloured title in addition to the crossed 'eye' symbol.

4.1.8 Restricting Size and Number of Files for Coursework, and Monitoring the Overall Size of Lectures Uploaded

Another setting that would need adjusting is restricting the number and size of files for submitted coursework. This can be done individually by selecting the 'edit settings' option through the editing function of the desired/respective coursework from the Course Front Page (see Figure 3.5). The option can be managed through the 'Submission types' block as illustrated in Figure 4.6.

Restricting the size of the submitted file(s) is important when there is an overall restriction on the space available for the programme (these restrictions are also discussed in Chapter 5 – Moodlecloud). The number of the submitted files should be set to 1 unless the task of the coursework requires more files – i.e., an Excel spreadsheet with calculations which has to be submitted additionally to the main (.pdf or .doc) file. This restriction will make marking easier, as the presence of more than one file by one

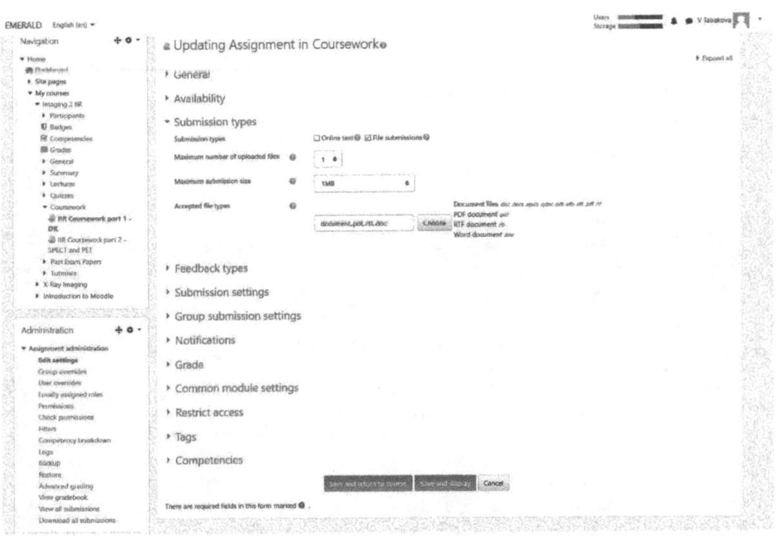

FIGURE 4.6   Edit settings for coursework – submission types.

student can lead to confusion if not properly named (the naming of the coursework files should typically adhere to the following requirements: the file name should include the title of the coursework in an agreed abbreviated style PLUS an ID number, allocated by the educating institution such as a student number – and not the name of the student for the obvious reasons of keeping marking anonymous).

4.1.9 Setting of the Grade Book

Another function for the Manager is the setting of the Grade book. This again has to comply with the requirements of the course – and more precisely with the course specifications. Moodle caters for different settings of the grade type (i.e., points, scale) and for our example the grading method would be 'simple direct grading' to a maximum of 100 points (Figure 4.7). The settings are performed from the front page of the course, in the section Coursework, for each of the two sub-sections: IIR Coursework part 1 – DR and IIR Coursework part 2 – SPECT and PET.

FIGURE 4.7   Setting the grade for coursework.

4.1.10 Weight Setting of Coursework

The two sets of coursework have different weights in the overall mark for coursework – in our example the Diagnostic Radiology (DR) coursework set has 35% and the SPECT and PET coursework set – 65%. The weight management is performed through the Administration block on the front page of the course – Course administration – Gradebook set-up. The weights are set manually in the fields provided in Figure 4.8.

4.1.11 Enrolment

There are also some functions that are not clearly 'Manager' functions and would more often be performed on the level of the Education Institution VLE Administration, such as a dedicated IT team. For instance, in the current version of Moodle on King's College London website, the Moodle VLE is highly individualised and bears its own name KEATS (an acronym for

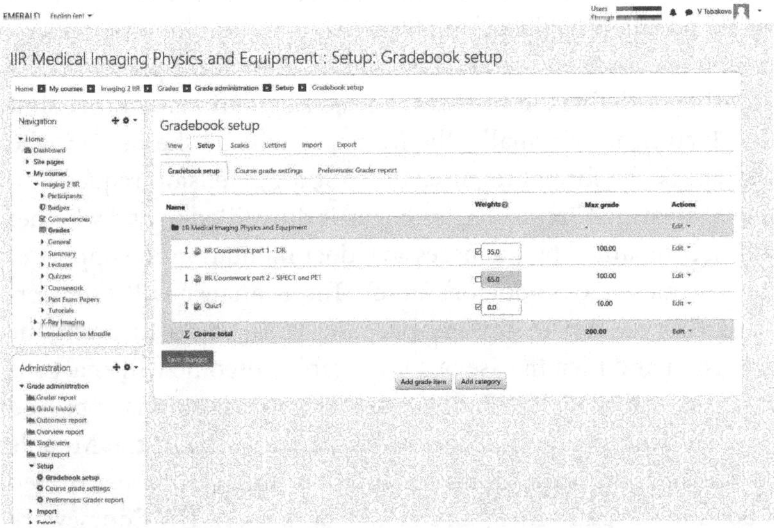

FIGURE 4.8 Setting the weight of a coursework grade.

King's e-Learning and Teaching Service and also a homage to the famous eponymous poet who studied there). It is synchronised with the administration e-platform of the university in a way that enrolment of students on the VLE is performed automatically if they have fulfilled a list of institutional requirements such as payment of fees and attainment of grades. However, even in a similar level of institutional involvement, the Manager of the course will be required from time to time to enrol different categories of users on the VLE. Other possible cases will involve the need to change the language. These and some other high-level functions are described in detail in Chapter 5, Section 5.2.

4.1.12 Altering Date of Submission

The Manager can also alter the date of submission of coursework for individual students after an application for extending submission has been received and approved by the authorised person or panel. This is necessary because after the cut-off date and time is set, submissions of coursework will not normally be accepted by the site after this date, unless an extension is given. The Manager would normally be the person receiving a request for an extension from the student and should be acquainted with the procedure for granting it (defined in the Programme Regulations or other similar document). Normally, the Manager will be the person who will check the formal requirements for an extension request – if the 'Mitigating Circumstances Form' is duly filled in and whether it is accompanied by the necessary documentation in support of their request (i.e., a doctor's note). The Manager will then forward this request to the person/panel authorised to grant the extension and then in case of a favourable outcome will enact the extension on the VLE. For instance, if an extension is approved for a student to submit Coursework part 1 – DR in the Moodle course Imaging with Ionising Radiation, this can be performed following the path: Imaging 2 IIR > Coursework IIR Coursework part 1 – DR > Grant extension.

4.1.13 Organising Feedback

The role of student feedback is valuable and very helpful to the educator. Furthermore, there may be institutional requirements for collecting feedback from students. If it is required that the Manager organises this activity, the steps to follow are illustrated in Chapter 3, Section 3.5.5.

4.2 THE ROLE OF THE STUDENT

4.2.1 Logging, Visibility, Profile Setting

The logging process for a student and setting up their profile will be similar to the actions described in Section 4.1. Once logged in, a student can view only content that is intended to be visible to that category of user. The most significant difference from the other users discussed (Teacher and Manager) is that there is no option for editing. In our example the only visible course to students is IIR Medical Imaging and correspondingly after log-in a student will view the following screen (Figure 4.9).

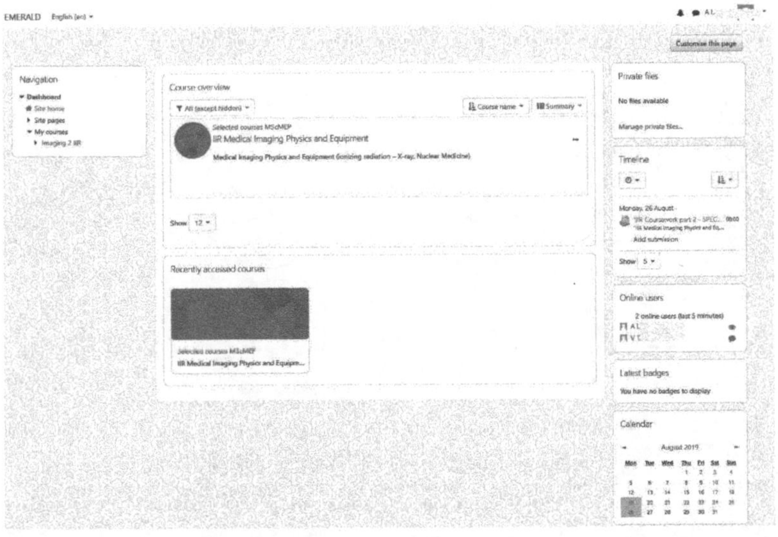

FIGURE 4.9   After login – student view.

N.B. In Figure 4.9 and elsewhere, students' names and e-mails have been masked for obvious reasons.

When selecting this course, the front page of the IIR course MEP will appear to the student in a version illustrated in Figure 4.10.

As compared to the front-page view of the Teacher (Figure 3.2), the folder 'Lecture archives' is not visible at all to the student. In the section Courseworks, the student has access to both sets of coursework and is required to submit them for marking.

The student is provided with information about his/her own submitted coursework. Figure 4.11 illustrates the screen bearing information of an already-submitted coursework (reached after selecting the respective title of the coursework from the front page); feedback and grade will appear when available. This view can be compared with Figures 4.18 and 4.19 in Section 4.3. There the same task has been completed and coursework has been submitted by two students and the Teacher can view the two submissions – and grade respectively – while the student can view only their own submission.

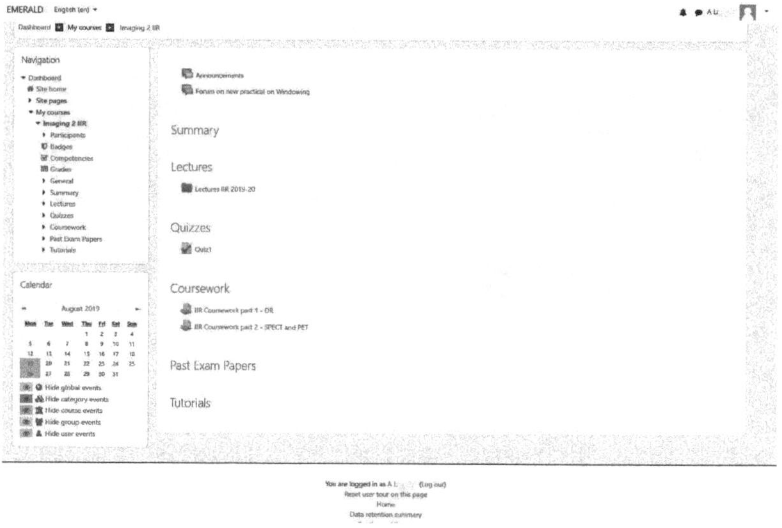

FIGURE 4.10   Front-page IIR – student view.

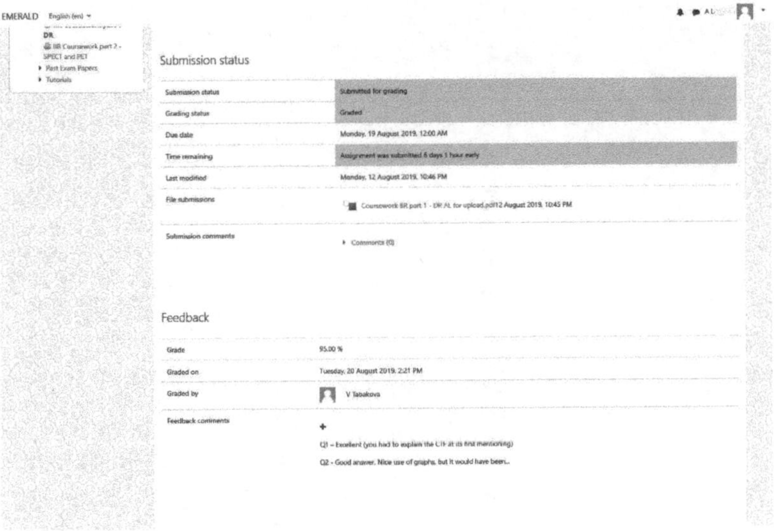

FIGURE 4.11   IIR coursework part 1 – DR submitted – student view.

4.2.2 Submitting a Coursework

The process of submitting a coursework is illustrated in Figure 4.12. This is accessed via Course Front Page, selecting the submission tool (a symbol of a hand holding a file) for the respective coursework – in the example 'IIR Coursework part 2 – SPECT and PET'.

After selecting the 'Add submission' button at the bottom of Figure 4.12, the screen shown in Figure 4.13 appears.

The submission for this type of Coursework is as a file attachment; in the example the submission is restricted to one file of the size of 1 MB.

It should be noted here that the Student agrees with the wording of the Declaration of own work (generally known as Anti-Plagiarism Declaration) by clicking on the submission button. There are various ways of dealing with the issue of plagiarism and they should be decided in conjunction with the Rules and Regulations of each educational institution. One of the widespread methods for conducting plagiarism tests on any submitted coursework is, for instance, the plagiarism detecting

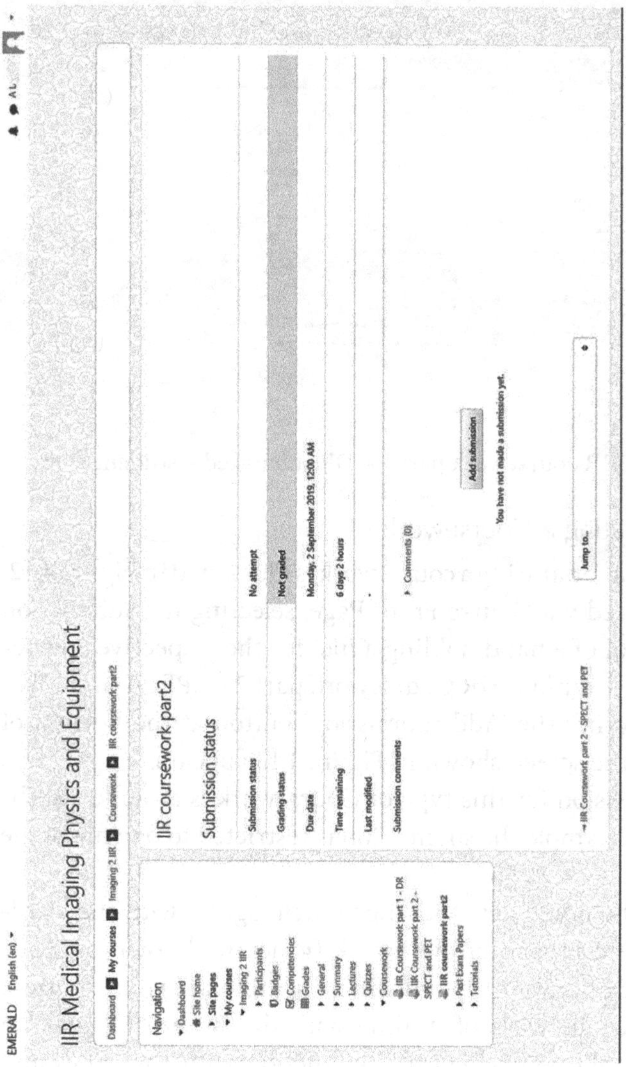

FIGURE 4.12 Submitting a coursework, step 1 of 2.

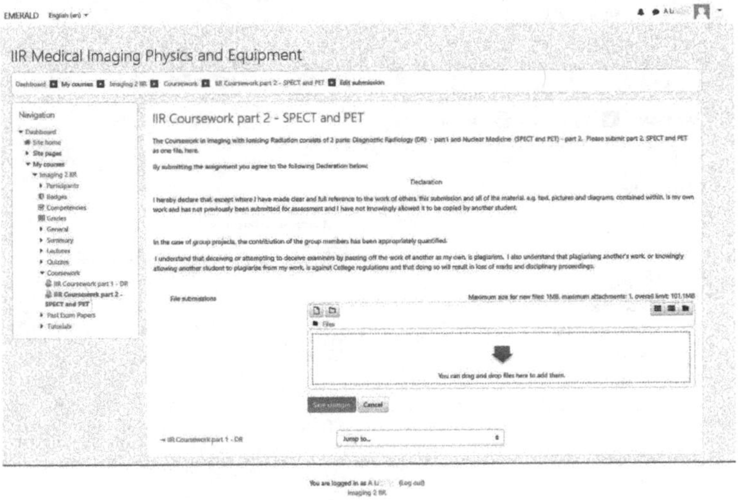

FIGURE 4.13   Submitting a coursework, step 2 of 2.

service TurnitIn. It can be embedded in Moodle and other VLE systems and is widespread in Higher Education. Independently of what method for detecting plagiarism is used, the implications of not conforming to the plagiarism requirements should be stated in the Programme Regulations and the Programme Director and Manager should draw students' attention to this at the very beginning of the Programme.

As in all previous actions for all types of users, for the changes to take place, they have to be saved using the dedicated button 'Save changes' at the bottom of the submission page.

4.2.3 Participation in Quizzes and Various Types of Communication

Another function in Student view is participation in quizzes and other types of assessment, as well as participating in forums and chats. Figure 4.14 shows the first section, General, on the front page of the IIR course containing the sub-sections 'Announcements' and 'Forum on new practical on Windowing'.

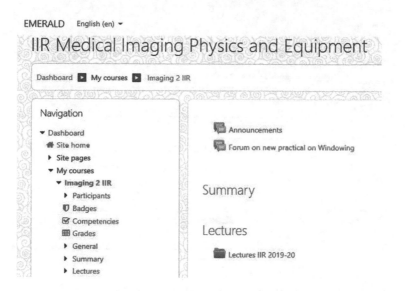

EMERALD English (en) ▼

IIR Medical Imaging Physics and Equipment

Dashboard ▶ My courses ▶ Imaging 2 IIR

Navigation

▼ Dashboard
  ▮ Site home
  ▶ Site pages
  ▼ My courses
    ▼ Imaging 2 IIR
      ▶ Participants
      🛡 Badges
      ☑ Competencies
      ⊞ Grades
      ▶ General
      ▶ Summary
      ▶ Lectures

📧 Announcements
📧 Forum on new practical on Windowing

Summary

Lectures

📁 Lectures IIR 2019-20

FIGURE 4.14   Front-page IIR zoomed General section; – student view with announcements and forum.

The student can participate in the Forum by selecting its title and then starting a new discussion topic within it (Figure 4.15) related to the forum.

In the example, the student starts the discussion topic by naming the subject: 'Useful new Practical' and in an attachment describes how the new Practical on windowing helped his/her work placement. The student also prompts others to share their experience in the Message section.

At the end the student selects the button 'Post to Forum' to enable the topic to appear within the Forum and the result is shown in Figure 4.16.

In other words, while a student cannot start a new Forum, they have the opportunity to add new discussion topics within an already created Forum. It is the task of the Manager, the Teacher, or in general the person with editing rights who has started the Forum to monitor for deviations from the main topic of the Forum. Note: A person with editing rights will have a slightly different view of the Forum (Figure 4.17).

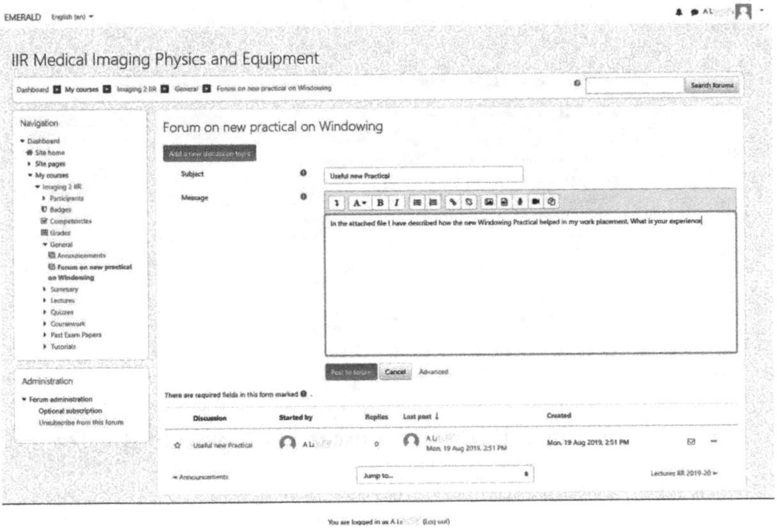

FIGURE 4.15   Title and description of the new discussion topic within the forum.

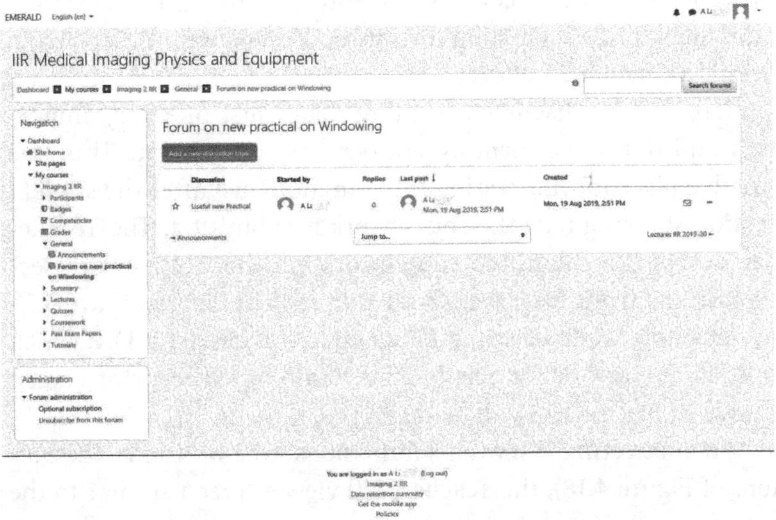

FIGURE 4.16   Result of adding a new discussion topic - student view.

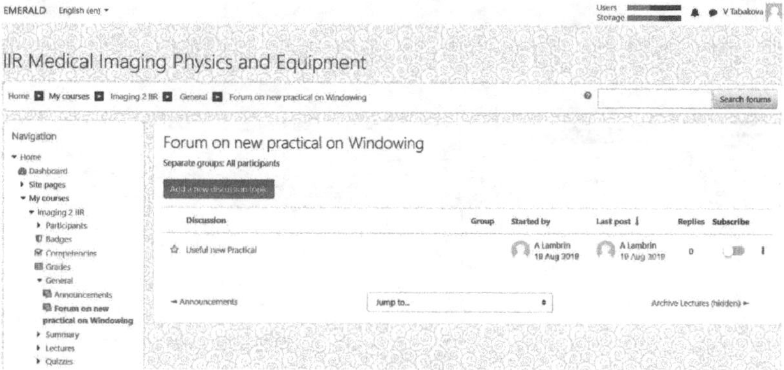

FIGURE 4.17 Forum with the new discussion topic added - Teacher view.

4.3 TEACHER FUNCTIONS: COURSEWORK ASSESSMENT

The Teacher functions related to building content on the VLE in the form of lectures, creating coursework and other types of assessment, viewing coursework submissions and performing assessment online, creating forums and chats, were described in detail in Chapter 3 before.

Here will be discussed in more detail just one very important addition – performing coursework assessment. This is a suitable place for illustrating this function, just after the student section showing how the coursework is submitted. The Teacher can access the submitted coursework via the Coursework section on the front page (i.e., from a screen similar to Figure 3.2). For example, after selecting IIR Coursework part 1 – DR, which is the coursework for Diagnostic Radiology, the print screen shown in Figure 4.18 will be displayed.

When selecting 'View all submissions' (the button at the bottom of Figure 4.18), the Teacher will view a screen similar to the one shown in Figure 4.19.

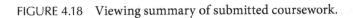

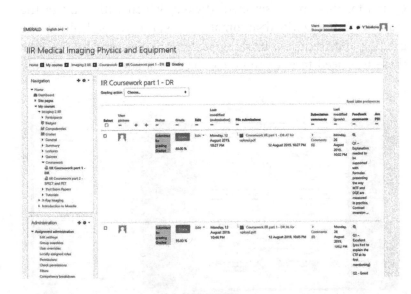

FIGURE 4.18 Viewing summary of submitted coursework.

FIGURE 4.19 Viewing all submissions.

If feedback and grading have not been performed yet, the columns 'Grade' and 'Feedback comments' would be empty. The process of giving feedback and grading can be performed online or offline.

4.3.1 Grading Online

The process is illustrated in Figure 4.20. (This figure appears when selecting the Grade button on Figure 4.19 for each submission.) The screen shows the submitted coursework to the left, the whole text of which can be accessed by scrolling up/down; on the right-hand side are the Grade and Feedback blocks, which are to be filled by the Teacher. The feedback given in the right-hand block refers to the whole coursework and not to a selected part of the submission.

A word of WARNING here: when performing the assessment, the Teacher should be aware of the box 'Notify students' next to the Save button (shown at the bottom of Figure 4.20). The default option for this box is 'selected' (ticked) which will result in the students being notified automatically via an e-mail about the grading and feedback. It is advisable to deselect the box in case there is a need to alter some feedback at a later stage and only

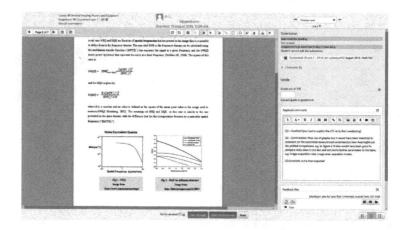

FIGURE 4.20   Grading online.

after all marking for all students is performed, to send an e-mail outside the VLE – and not to use this automatic notification. It is best to send a group e-mail on behalf of the Programme Director to notify students to log in to the VLE to view their assessment.

Note: In the example shown in Figure 4.19, the submitted assignments are two and they can both be viewed by the Teacher; compare this with the Student view shown in Figure 4.11, where the Student can view just his/her own assignment.

4.3.2 Grading Offline

By selecting the button for 'Grading action' at the top of Figure 4.19, a drop-down menu appears as illustrated in Figure 4.21. The Teacher can then select to download all assignments (the first option from the drop-down menu). As a result of this the Teacher will be prompted to open or save the submissions as a zipped file. The assessment can then be performed offline and the Teacher can upload marks and feedback at a later time.

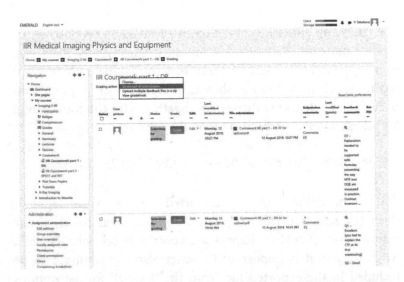

FIGURE 4.21 Choosing the grading action to download all submissions – teacher view.

4.4 MOODLE AS A SOURCE OF SPECIFIC PARTICIPANT INFORMATION (GRADES; ACTIVITY): THE MANAGER/TEACHER VIEW

As a source of information, Moodle presents broad opportunities. One of them is an overall view of the grades for the participants in a given course. Figure 4.22 illustrates a grader report featuring the two students enrolled on the course, each one of them having submitted two sets of coursework and one quiz. In the case of a greater number of students, a search can be performed by the first and last names through the alphabetical selector appearing at the upper part of Figure 4.22.

It is worth noting here that due to previous weight settings (Section 4.1.10), the mark for the course total is not the arithmetic mean and is rather calculated by a formula. The formula is set up typically by the Manager in accordance with the course specifications and, in our example, is set up to the formula:

$$35\%[\text{CW1 mark}] + 65\%[\text{CW2 mark}] + 0\%[\text{Quiz mark}].$$

The setting up of the weight is explained earlier in Section 4.1.10.

For a more detailed view of all users' grades and feedback, the user report in Figure 4.22 (the last from the list in Grader report) has to be selected. Then as shown in Figure 4.23 all grades for one of the students are displayed, together with the course total for all assessments, which is formed by the equation:

$$95 \times 35\% + 75 \times 65\% + 100 \times 0\% = 33.25 + 48.75 + 0 = 82\%.$$

This information can be exported to an Excel spreadsheet (by selecting from the left Administration panel 'Grade Administration'>'User Report'>'Export'>'Excel spreadsheet'). At this stage, it is important to select also the feedback to be included in the exported file from the 'Export' format options (Figure 4.24).

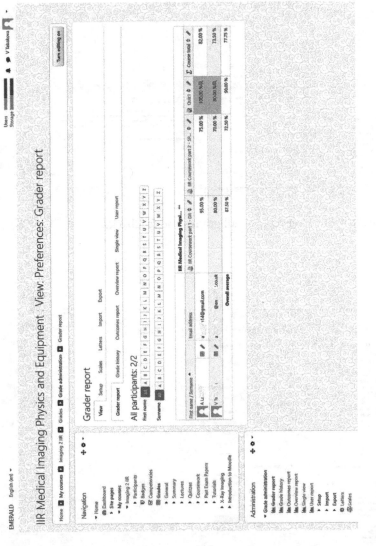

FIGURE 4.22   Grader report – all students.

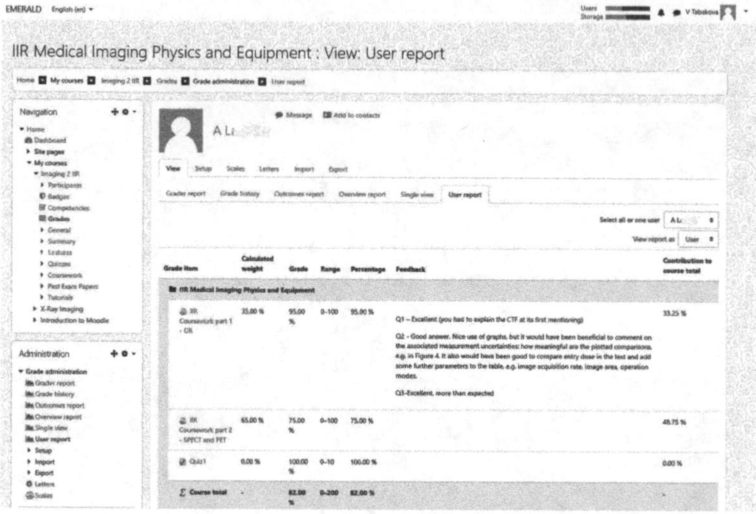

FIGURE 4.23   User report – all grades and feedback.

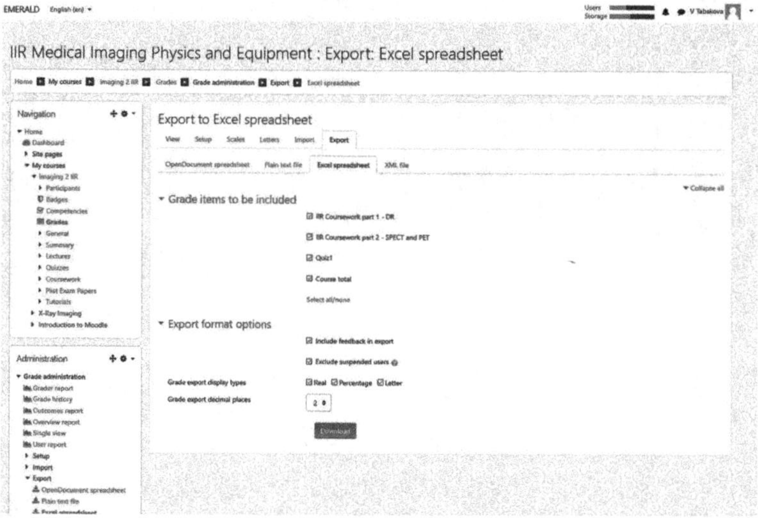

FIGURE 4.24   Exporting grading information.

This is an important step useful for archiving student information (i.e., required for upcoming Exam Boards; for all types of quality control on an institutional level and for outside national and international accreditation, as well as for safeguarding against data loss and for general backup). It is advisable that this action be performed at least once at the end of each course.

Moodle can also be used for information about student (and all user) activity. The resulting outcome is shown in Figure 4.25. The information is reached via Course Administration – Reports – Logs. Please note that the IP address (hidden here) will also appear in the report.

This information is very useful for monitoring of students at periods when they are not present on the campus of the university (such as in the case of Medical Physics when they are in their work placements, but are still required to do some academic work). In one such case at a university, the VLE Manager

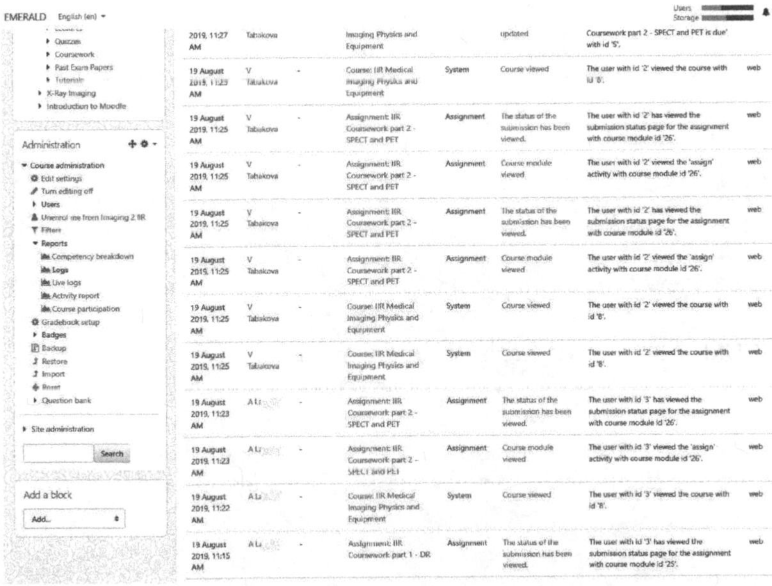

FIGURE 4.25 All user activity – personalised.

of a part-time MSc Programme in MEP noticed that a student had not been active on the VLE for several weeks. The local supervisors were contacted and it was revealed that the student had gone into long-term sick leave; the necessary steps were undertaken to organise an extension year for the academic studies at the university for the student.

The activity reports (Figure 4.26) can be used as information about the most popular parts/pages of the course visited by the users (this is anonymised information in the example) and that information can be analysed further.

FIGURE 4.26   Activity reports – anonymised.

CHAPTER 5

Aspects of Moodle Application

5.1 e-LEARNING IN MEDICAL PHYSICS IN LOWER- AND MIDDLE-INCOME (LMI) COUNTRIES: RESULTS FROM A SURVEY

In order to encourage the use of e-Learning in the profession, the author presented a guide to organising a Virtual Learning Environment (VLE) platform at several international forums addressing mainly representatives of the profession of Medical Physicists and Medical/Clinical Engineers from LMI countries, namely:

- IOMP School session at the International Conference on Medical Physics, Bangkok, December 2016;

- The 17th Asia Oceania Congress of Medical Physics (AOCMP), Jaipur, November 2017; (special session)

- The World Congress on Medical Physics and Biomedical Engineering, Prague, June 2018

- The College on Medical Physics: Applied Physics of Contemporary Medical Imaging – Expanding Utilization in Developing Countries at the International Centre for Theoretical Physics (ICTP), Trieste, August-September 2018; (special session)

- and for broader applications of VLE in the medical profession before academics at the Plovdiv Medical University, June 2015; (special session)

The attendees at those lectures amounted to approximately 100 professionals cumulatively who were directly involved in leading positions in teaching and academia in Medical Physics and Engineering in various LMI countries. The questionnaire distributed to the participants is shown in Table 5.1.

TABLE 5.1   Questionnaire e-Learning in LMI Countries

Computers available for staff and/or students at your university/ institution	Less than 10%				
	10%–30%				
	30%–50%				
	50%–80%				
	Above 80%				
	Don't know				
	N/A				
Computers used by students at home in your university/area	Less than 10%				
	10%–30%				
	30%–50%				
	50%–80%				
	Above 80%				
	Don't know				
	N/A				
Availability and stability of internet in your area/ country (1 – low; 5 – high)	1	2	3	4	5
Do you have personal experience in using e-Learning?	Yes, as a student				
	Yes, as a lecturer				
	Yes, as both				
	Neither				

(*Continued*)

TABLE 5.1 (*Continued*)   Questionnaire e-Learning in LMI Countries

If you are using e-Learning which platform do you use?	Moodle
	WebCT
	Blackboard
	Other (please list here:.........................)
What advantages do you find in your e-Learning platform? (List as many as applicable)	Facilitate delivery of lectures
	Higher lecture quality (i.e., high-quality images)
	Facilitate assessment (incl. delivery of coursework, introduction of quizzes, and self-assessment)
	Facilitate update of material
	Facilitate communication with students
	Possibility to increase intake (number of students)
	Increase student satisfaction
	Save time
	Other (please list here...............)
	I don't use e-Learning for teaching
Do you plan to use e-Learning for your activities in the coming 1–2 years?	Yes, as a student
	Yes, as a lecturer
	Neither
	I already use e-Learning
What reasons would prevent you/your institution to implement e-Learning in the coming 1–2 years? (Tick as many as applicable)	Lack of technology (computers, low speed of internet)
	Lack of finance
	Lack of trained IT staff
	Lack of experience of lecturers with e-Learning
	Lack of commitment from top management
	Lack of suitable lecture material

(*Continued*)

TABLE 5.1 (*Continued*)   Questionnaire e-Learning in LMI Countries

	Time to prepare material suitable for e-Learning		
	Other (please list here......................)		
Please enter your country			
Please enter your institution	Hospital	University	Other

Of all the attendees, 79 submitted a filled-in questionnaire. The analysis of the answers showed the following:

- Computers available for staff and/or students at your university/institution – 60% of participants reported that the availability of computers at the institution is 50% or higher.

- The use of computers at home varied from 15% to 60%.

- The stability of internet understandably varied from country to country, but was overall satisfactory (rated 3) and above.

- 28% of participants declared that they do not have personal experience in using e-Learning – neither as students nor as lecturers.

- Of those who had some experience in e-Learning, the most common platform in use was Moodle (72%). Other platforms were reported, mainly by those who had received their education in the USA; i.e., Canvas, WebCT, Blackboard – 17%, and non-reported – 11%.

- Of the advantages of e-Learning platforms, the most highly valued was the possibility to update the material (the reasons behind this are the dynamics of the profession of Medical Physics and Engineering, and the high volume of images, the updating of which is facilitated enormously by an e-Learning environment). Other advantages highly ranked are as follows: saving time, increased student satisfaction, facilitating assessment, and communication with students.

- About 5% were already using e-Learning platforms in their teaching activities.

- The professionals intending to use e-Learning for teaching activities in the near future was 34% at the beginning of the lecture; it rose to 65% after the lecture.

The broader analysis to the effect of this question is very interesting as the attendees at these International Symposia and Congresses were not limited only to participants from LMI countries. Several participants from countries with a very long tradition in e-Learning such as the USA, Australia, and Japan also attended the author's presentations. They provided oral feedback after the presentations reporting that they feel more confident now to start using e-Learning in their teaching. (They did not fill in the questionnaire on Table 5.1.)

- The main reasons that would prevent the attendees to implement e-Learning in the near future were

  - Lack of experience of lecturers with e-Learning (80% of all answers).

  - Lack of suitable teaching material – 68%.

  - Lack of finance – 78%.

  - Lack of trained IT staff – 77%.

  - Technical problems – i.e., low reliability of internet – 19%. Although this is not a high percentage, it is necessary to bear in mind that for countries without reliable internet, establishing e-Learning on a VLE platform is not viable and alternative ways of delivering the knowledge should be sought, such as making educational materials available for download. A very suitable package for educational programmes in Medical Physics in such case would be the EMERALD materials which have been distributed free of charge worldwide since 2002. For further information, visit www.emerald2.eu. The information on the site is suitable for entry-level knowledge (at least BSc) in Medical Physics.

The questions about the intention to implement e-learning and the main reasons preventing professionals to embrace e-Learning, according to Table 5.1, were placed before 22 Medical Physics professionals at ICTP twice, in September and December 2018. The second survey was conducted after a 2-hour lecture and two 1-hour hands-on sessions on Moodle.

5.1.1 Overall Outcome

The results from the second survey showed that the participants felt more willing and confident of engaging with e-Learning platforms and the number of those rose, with an increase of 75%. Parallel to this, attendees reported that they are now quite clear about the steps for preparation of teaching materials to fit the requirements of e-Learning platforms and they are also more aware of the financial and IT requirements for introducing e-Learning and feel that these hurdles can be overcome.

5.2 THE MOODLE CLOUD: STARTING FROM SCRATCH

The results from the surveys clearly show that Medical Engineering and Physics professionals in LMI countries have the will to start/ broaden their use of e-Learning as educators. This is strengthened by the fact that until 2035 the need for Medical Physics professionals worldwide will need to be almost tripled [1,2]. e-Learning is a powerful tool to help solve the requirement for the three-fold increase in the number of professionals. And realistically a free and open source VLE platform such as Moodle will be very appealing to countries with restricted resources. However, there are still accompanying costs involved which need be taken into consideration. These are discussed below in the different scenarios that may be encountered by educators.

5.2.1 Choosing a Host Server

5.2.1.1 Option 1: *Installing Moodle on Own Server*

Moodle is completely free if the user installs the software on their own server. However, in this case costs for maintaining the server as well as technical staff (IT) to maintain the software will be necessary. In our experience, the dedicated IT staff for a programme such as Medical Physics need not be responsible full time; the programme would take about 0.25 of the time of one IT person to dedicate to such a programme at the beginning and will be down to 0.1 of this when the programme is established. So within an educational institution, it makes sense that one IT person deals with 4–10 programmes on the VLE. This option is viable when there is a combination of several programmes to share the costs for IT personnel and server, and a decision need to be taken at the level of the management of the educational institution. Instructions on installing Moodle are available at https://docs.moodle.org/en. Complete installation packages are available for Windows and Mac OsX. The link above also provides instructions to set up one's server.

5.2.1.2 Option 2: *Limited Budget and No Server*

Another option for educators with limited budget and no server is to take opportunity of Moodle's offer to run the Learning Management System on its own server (MoodleCloud). There is a possibility here to use Moodle's server for free (the main constraints are the number of participants on the VLE and the desired volume of file storage).

This option is suitable also for a free trial.

The first step when using a MoodleCloud server is for the person responsible for organising the educational programme to create a new account on https://moodlecloud.com. This process is rather straightforward and requires agreement to the terms of service and entering some personal information (name, e-mail, phone, institution).

5.2.2 Acquiring the Space for the Programme, Log-in, and Initial Customisation

As a result, the educator acquires a new, empty Moodle site (for free and on MoodleCloud server). His/her role in MoodleCloud is called 'Administrator', and it is the role with the highest level of functionality, resource access, and management for the site.

N.B. The literature on Moodle which is available on the internet and sites like Amazon often contains a reference to Administrators (very often already in their titles). Such literature is designed to be relevant for the skills of System Administrators with specialist IT knowledge. For building a course on MoodleCloud it is not necessary to have this type of knowledge.

For the option chosen – free hosting on MoodleCloud – the number of Administrators for the site is limited to one. The Administrator first creates the space for the new programme on MoodleCloud. Then s/he logs-in on the portal of the already created site, and will be prompted to personalise it initially by entering a name for the website. When this is performed, the URL of the site will be https://yourchosenname.moodlecloud. com. Following the steps described above the author built new sample site for MSc Medical Engineering and Physics and named it 'EMERALD'. In this way the URL of the sample site is https://emerald.moodlecloud.com

The login for the Administrator by default will be with username: **admin**.

The whole structure of the site will depend on the Programme Specifications, but before coming to that point, there are several steps to be taken into account.

5.2.3 Choosing the Theme of Appearance

There are two pre-set themes of appearance at the time of writing, 2019. These are Boost theme and Classic theme; the default one being the Boost theme. Both themes work well on mobile devices as well as desktops, and may be customised from the 'Themes' area of the block 'Site administration'. (For illustration of the block 'Site administration,' see Figures 3.5 and 3.6, left panel.)

To switch between the two pre-set themes, choose Site administration – Appearance – Themes. In Appearance for HTML settings use Default ones in each box. In this way the e-Administrator of the VLE need not be an IT person and no knowledge of JavaScript or any other programming language is required.

5.2.4 Structuring the Site

There will be just one course appearing on the front page – the Introduction to Moodle. The page is virtually blank for the Administrator to start introducing new courses as described in Chapter 4.

All changes are introduced by selecting first 'Edit settings' from the Administration block on the front page. (For reference on the block, see Figure 3.5, left panel.)

In this way the full name and short name of a course can be changed. (The short name appears in the navigation bar.)

Then the Administrator will decide what should be shown on the front page – news items, courses, course categories, etc.; will these be the same or different for logged in and non-logged in users? As with other settings, they can always be changed later.

In fact, all other steps for creating a Programme with several courses on Moodle have already been described in Chapters 3 and 4; here is a summary of all these steps:

- using the editing function;

- adding a new course;

- creating content – lectures;

- creating content – other (audio files, video files, URL);

- creating content – coursework;

- creating content – other assessment (i.e. quizzes);

- increasing student involvement – creating forums and chats;

- performing assessment – online and offline.

One of the most important steps remaining to be discussed is enrolling users.

5.2.5 Enrolling Users
It has to be noted that this is a two-step process.

5.2.5.1 Authentication
This is the first step in the two-tier process.

The function is reached in Site administration – Users – Add a new User. (For reference to the block 'Site administration', see Figure 3.5, left panel.) The required fields are First name, Surname, e-mail address, and User name (Figure 5.1). New users (students and Teachers) are added either individually or by creating accounts via a file, storing tabulated data for the students to be enrolled (such as a file with a csv or excel format). For most MSc MEP courses in LMI countries the number of students on any given MSc course in MEP won't be large and therefore new users can be entered individually.

At this stage, it is not necessary (and is not desirable) to assign a role to the user, as one and the same person may be required to have different roles in different courses. Once created, the authentication allows to go to the second step.

5.2.5.2 Enrolment
Once users have an authenticated account, they can be enrolled in courses. The time to assign a role to them (student, teacher, guest) is the moment a user is enrolled on a specific course.

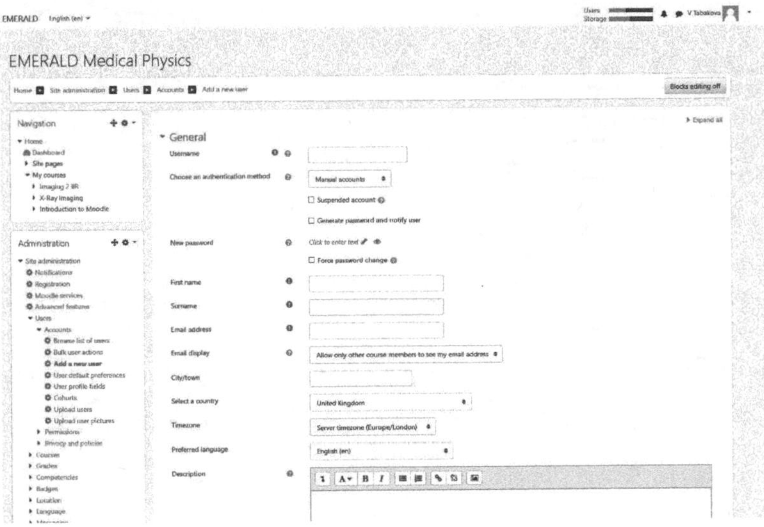

FIGURE 5.1   Adding a new user – authentication.

Although there is a possibility for self-enrolment, this, in our experience, was not normally used as the educational institution will need to check if all the necessary prerequisites for enrolment have been fulfilled (i.e., fees paid, certain grades achieved, etc.). Again, for courses in the discipline of Medical Engineering and Physics, especially in LMI countries where the number of students won't be big, it is viable to enrol students either manually or via a file containing tabulated information.

For enrolment to a particular course, it is necessary to go to the front page of the relevant course (i.e., on Figure 3.5) and in the Block 'Administration' (left panel on Figure 3.5) to select Course Administration > Users > Enrolled Users. The following screen appears (Figure 5.2).

By selecting the button 'Enrol users' at the top right corner of the 'Participants' block the Administrator is prompted to select a user from a list of already authenticated user accounts (Figure 5.3) and to assign a role to the selected user (in this case a student role is assigned to user *gs18* – Figure 5.4).

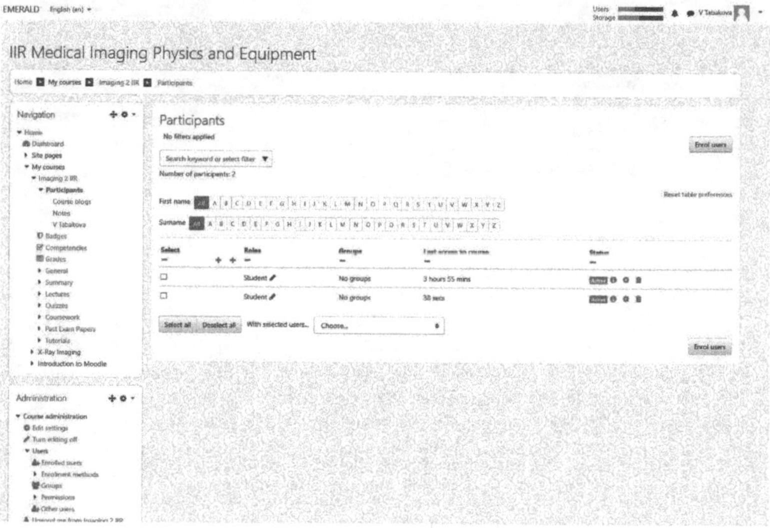

FIGURE 5.2   Summary of enrolled users.

The final step is to select the button – Enrol users. Following these steps several users from the list shown in Figure 5.3 can be enrolled.

If necessary, the students already enrolled in a course can be divided in groups or allowed to view and participate in all activities as a whole, undivided group.

5.2.6 Selecting a Language

For a globally significant profession such as Medical Engineering and Physics, it will be necessary for some geographical areas to set the preferred language of instruction. This is performed through Site administration – Language (Figure 5.5).

There is currently a selection of over 130 options for a language in all major alphabets the varieties of a language are broadly catered for – for example, there are separate versions for German and Swiss German; French and French (Canadian), five versions of Spanish; for many languages there is also a workplace version, with the terminology better suited to a workplace and not an

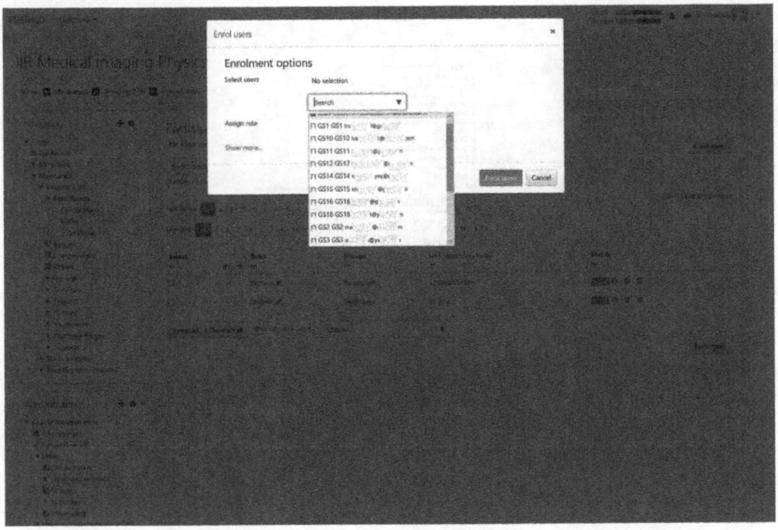

FIGURE 5.3   Selecting from a list of authenticated accounts.

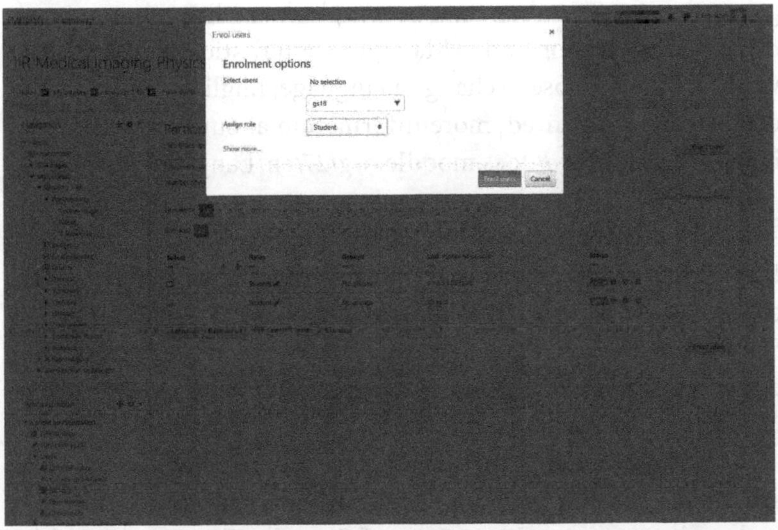

FIGURE 5.4   Enrolling a new user to a course and specifying the role.

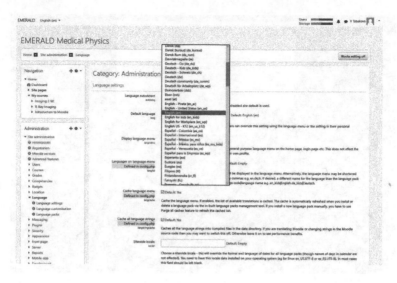

FIGURE 5.5   Selecting a language.

educational institution (for instance Teacher would be 'Trainer' and Student – 'Participant'). There are also 'child' language packs, which contain only the modified language strings from their parent language, rather than a complete set. For our purposes the e-Administrator and e-Manager should stick to the Classic version if they choose to change a language; English (en) being the default one. If required, more information about language packs is available at https://docs.moodle.org/27/en/Language_packs.

5.2.7 Title Role Management

If the titles of the roles need to be changed for some reason, this can be performed by the action of 'managing roles', reached through Site administration – Users – Permissions. (Figure 5.6.) The default options for the role titles are: Manager, Course Creator, Teacher, Non-editing teacher, Student, Guest, etc.

Apart from moving the roles up and down using the adjacent arrows to each title, a more important opportunity for our case is the editing of the short name of the role. This is achieved by clicking on the 'edit' (cogwheel) button to the right of each role.

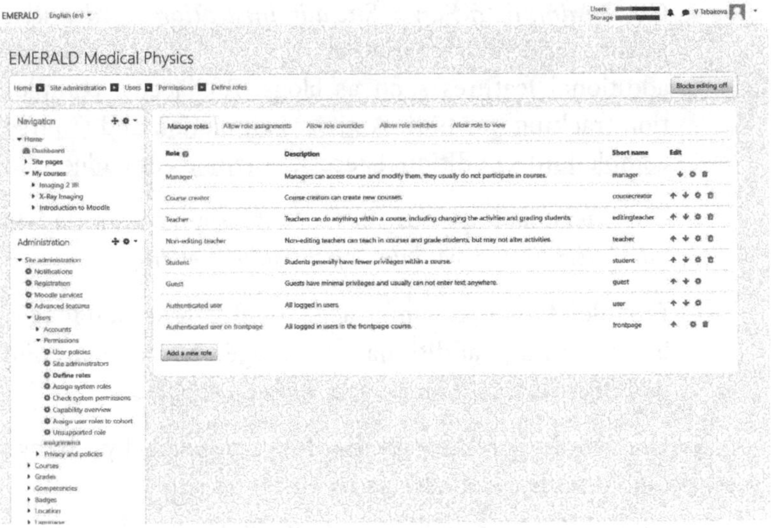

FIGURE 5.6   Managing roles.

However, it is not advisable to change the short name of a standard role (i.e., Manager, Teacher, Student).

5.2.8 Other Tips for Quick Start

Moodle also provides tips for quick start for teachers at (https:// docs.moodle.org/19/en/Teaching_tips_and_tricks), of which the most useful for our purposes will be performing back up and importing course data. Tips for automatic course enrolment may be useful for larger courses (over 10–15 students).

In addition, Moodle provides the following tips for Administrators:

- Go through each activity in *Administration* > *Site administration* > *Plugins* > *Activity modules* and decide the most suitable default settings for your Moodle. Do the same for the gradebook, via *Administration* > *Site administration* > *Grades* > *General settings.*

- Go to *Administration* > *Site administration* > *Advanced features* and review whether you want to make use of additional features such as blogs, RSS feeds, completion tracking (for both Activity completion and Course completion), conditional access, portfolios or badges.

- Consider enabling your site for Mobile app access via *Site administration* > *Mobile app* > *Mobile settings*

- Provide the user interface in different languages by installing additional language packs via *Site administration* > *Language* > *Language packs.*

- Use wording more suitable to your users by changing Moodle's default terms in *Administration* > *Site administration* > *Language* > *Language customisation.*

Some of these options are not always applicable for our case – especially changing the short names of a standard role.

An example of highly customised educational site on Moodle can also be viewed at http://school.demo.moodle.net, where a demo, hypothetical education unit called 'Mount Orange School' is created.

The site has been designed for anyone to see and play with Moodle 2.0 and explore ways of using Moodle in places where people teach and learn.

The roles to explore are student, teacher, parent, and privacy officer at Mt Orange School.

5.3 DISCUSSION

We have discussed Moodle until now on the basis of its features of being free and open source. The term 'open source' is easily defined as the code is supplied for free by the developers and allows for a variety of customisation. The 'free' feature needs to be explained in more detail. The first question before educators having decided to implement a new VLE is how to get started?

For proprietary VLEs supplying their platform on paid terms the answer is straightforward – contacting the supplier and getting a quote. With Moodle there are more decisions to be made upfront.

The first and most important of these is: Do you want to install Moodle on your own server or would rather have it hosted externally? And what exactly is free?

The free ingredients in all options discussed below are:

- The Moodle software is free to download.

- The Moodle code is available for free, that is the meaning of 'open source' – and this allows customisation.

- The Moodle upgrades are free, but one can opt out of them (if you prefer you can have your Moodle platform not participating in the upgrades).

So, the options before an educator who wants to implement Moodle are as follows.

5.3.1 Option 1: Installation on Your Own Server (Self-Hosted)

Install Moodle on your own server. It requires a web server with PHP and a database. *PHP* is defined as a recursive (repetitive) acronym for *PHP*: Hypertext Preprocessor and is a scripting language used to develop websites or web applications.

Moodle is open source under the GPL (General Public) licence. GPL is a free software license, which allows users to modify the software. The GPL requires that any work derived/based on the original should also be open-source and distributed under the same license terms.

Moodle states that 'Everything we produce is available for you to download and use for free'. Detailed information on the subject can be found on https://docs.moodle.org/37/en/Installation.

The most important features for this option are as follows:

- Free download.

- Unlimited users.

What do users have to budget for?

In this first option - using an own server finance will be required for:

- Server space – the cost for a small-medium sized university will be c.300 UK pounds per month for 2,000 students. This cost obviously will be shared with other educational programmes at the institution.

- Upfront cost for server – c.4000 USD [3].

- Training of personnel – IT staff and Administrator for the course.

- Maintenance – three to four personnel for all university of the above size for the maintenance of VLE for all programmes, in our experience the estimated time needed will be c.0.25 IT staff for Medical Physics, (or even less); in other words, an IT staff will be engaged with the Medical Physics programme maximum one day a week, including general maintenance on the server. Of course, this estimate is realistic when there is one dedicated full-time Administrator for the MSc programme, or if the Lead Teacher for every course on the programme engages in managing their area.

5.3.2 Option 2: Using Moodle's Server (MoodleCloud)

This option is free or paid, depending on the usage tier selected (see Table 5.2).

The main advantage when using this option is that the educator is getting a robust, reliable, and affordable hosting by Moodle. It is very suitable for small to medium organisations.

TABLE 5.2 Moodle Cloud Options – Using Moodle's Own Server

			Moodle for School		
	Free	Starter	Mini	Small	Medium
Annual price in AUD (as per April 2020)	$0	$80	$250	$500	$1,000
Max number of users	50	50	100	200	500
Max file storage	200 MB	200 MB	200 MB	400 MB	1 GB
Personalised site name	Yes	Yes	Yes	Yes	Yes
Mobile app enabled	Yes	Yes	Yes	Yes	Yes
Web conferencing with BigBlueButton	Yes	Yes	Yes	Yes	Yes
Max users per session	10	50	100	100	100
Session recording	x	x	Yes	Yes	Yes
Inactive site retention*	x	Yes	Yes	Yes	Yes
Custom certificates	x	Yes	Yes	Yes	Yes
Document converter	x	Yes	Yes	Yes	Yes
Automated backups	x	Yes	Yes	Yes	Yes
Advanced theme	x	x	Yes	Yes	Yes
Extra plugin pack**	x	x	Yes	Yes	Yes
Plugin and theme installation	x	x	x	X	x

Source: https://moodlecloud.com/app/en/
* A free site if inactive for 60 days will be removed. WARNING: As per March 2020, the inactivity period has been changed to 2 weeks.
** The extra plugin pack includes a plugin that allows an attendance log to be kept. Other plug-ins in this pack are related to the application of typical elements of game playing (e.g. point scoring, competition with others, rules of play) to encourage better student engagement and to enhance student experience, etc. More detailed information is to be found at https://moodle.org/plugins/browse.php?list=set&id=80.

In this option the VLE platform is hosted on Moodle's server – The Moodle Cloud. The main benefits of this option are connected to the fact that there is no need for the educator to be (or to hire) a platform developer.

5.3.2.1 MoodleCloud Packages

There are (as of 2019) three MoodleCloud packages: Moodle for Free, Starter, and Moodle for School (with three tiers/plans).

Each package contains a set of features, including plugins and one or more themes.

The Moodle for School package has three plans, each with a different user and data allowance, but otherwise they all have the same features. These packages with their cost in Australian Dollars (AUD) are given in Table 5.2.

Further details are available on Moodle.com/cloud.

If the aim is to start a single module or several modules from an MSc programme of Medical Physics on the VLE platform, then the free option or the starter option from Table 5.2 is suitable. WARNING: As of March 2020 a free site if inactive for 14 days will be removed from the MoodleCloud server.

5.3.2.2 Main Restrictions and Possible Ways to Deal with Them

The main challenge will not be the number of students, as the classes would normally be below 20 students in a medium-sized LMI country. However the number of students will affect the maximum file storage capacity, discussed below.

A major consideration here will be the number of teachers – for a class of 15 students, the number of teachers for a single course can reach 5–7 for a course like Radiotherapy, where each sub-topic would require a professional with a narrow specialisation. The number of teachers can well be in the region of 20–25 across all courses in the curriculum for a class of 10–15 students. The number of teachers need not be greatly increased if the number of students rises to 20–25. Above this number, there may be a necessity to include more assistant teachers specifically/predominantly to help with the assessment of coursework. Realistically, a number of combined users (students and teachers) exceeding 100 is unlikely to be needed.

The other restriction to have in mind will be the limit for maximum file storage.

The overall limit for the free option as well as for the next two levels currently priced at 80 and 250 AUD per annum, respectively is 200 MB. The limit refers to all lectures and all student submissions, so it is obvious that 200 MB will be reached quickly with lectures in two to three courses and five to six students uploading the relevant coursework, as Medical Physics is a subject matter with lots of imaging. In the example in Chapter 3 for the course 'Imaging with Ionising Radiation', which is a working example, the size of all image-rich lectures in PDF format is 70 MB and if a single lecture in Power Point Format PPT (a 25MB one in the example in Chapter 3) is added, the overall size will reach half of the allowance even without taking into consideration students' coursework to be uploaded (which will also contain a lot of images and diagrams).

It is obvious here that especially in this scenario it is necessary to impose restrictions on the size of submitted files for coursework, and in most cases in our experience the limit of one file and 1 MB are very reasonable. (How to set the restrictions is described in detail in Chapter 3.) This scenario requires very strict housekeeping, ensuring deletion of large files if they are not needed any more (for instance as discussed earlier in Chapter 3, with video or audio files relating to a pre-exam tutorial), and archiving student coursework outside of the VLE (i.e., in a dedicated space on the computer of the e-Administrator or Programme Director) after the assessment is performed.

It is possible to accommodate all courses in an MSc Programme in Medical Physics at the highest end of this option, at a cost of 1,000 AUD per annum and 1 GB overall limit but again very strict housekeeping needs to be implemented – with files not in use being taken down from the platform, so that the maximum file storage limit is not exceeded. Additionally, for audio and video files needed to be deleted when obsolete (discussed earlier in Chapter 3), attention should be paid to Research Projects and MSc theses if they are planned to be uploaded on the VLE platform. It may well appear that the overall limit is exceeded in case of MSc theses submitted on the VLE and then alternative solutions for submitting outside the platform should be planned. In any case, there should be a very profound cost-effectiveness analysis performed before a commitment is made, especially as there may be cheaper options like hosting the platform based on Moodle using a third-party service provider and not on a Moodle server.

This should include comparison with Option 1 (Section 5.3.1), and the result may well be that it is financially more viable to choose installation on own server, or even to choose server hosting from a third-party service provider.

5.3.3 Option 3: Using a Moodle Partner

There is a third option – setting up Moodle with a Moodle Partner. In this case, the number of users is unlimited and the server would be either acquired/owned by the education institution or hosted on a third-party service provider. A list of certified Moodle Partners is available on the Moodle website, and for the current list, visit https://moodle.com/partners/. These are certified Moodle experts, listed according to geographical region and sectors (Higher Education, Schools, Workplace). Services range from consulting to development, hosting, installation, customisation, support, and training. Again a cost-effectiveness analysis is necessary to choose the right option, taking into account various factors such as weighing the long-term benefit (and expenses) of educating in-house IT staff vs. using a Moodle Partner, the size of the educational institution where the MSc programme is delivered (for the possibility to share the server and all other expenses with other programmes) and quotes from certified Moodle Partners can help with the decision.

5.4 SOME COST AND OTHER CONSIDERATIONS ON SEVERAL PROPRIETARY (PAID) VLE PLATFORMS

5.4.1 Blackboard Learn

It is not possible to compare directly the overall costs of introducing Moodle with that of proprietary (paid) VLE platforms costs as the pricing of Blackboard Learn (previously known as Blackboard Learning Management system) is made known only after prospective users contact the company asking for a quote and further details. Here, it is important to bear in mind that Blackboard is most often purchased by larger educational institutions looking for a tailored VLE solution. Smaller institutions may find the information below useful in order to weigh all available options.

Information below is taken from discussion forums on the internet as an attempt to compare prices indirectly:

- "Pricing for Blackboard and Others?" Forum https://moodle.org/mod/forum/discuss.php?d=193196 – In 2012, a Moodle forum post asked for help pricing Blackboard with 300 learners. Most of the responses highlighted the benefits of Moodle, there was one response stating that one school was quoted at $1,200 (USD) per student per year for Blackboard, adding up to $36,000 USD for a class of 30 students. While this seems unrealistic, a more realistic entry can be found in the discussion forum "Blackboard by the Numbers"

- Blackboard by the Numbers On https://moodle.org/mod/forum/discuss.php?d=193196, a user estimates the average cost of a customer license to cover a university or school district with the full Blackboard package to be $160,000/year. This assessment is based on research from Instructional Media+Magic (im+m) [4] However, it should be taken into consideration that this article is from 2006.

Bottom Line: The only way to accurately estimate the cost of Blackboard is by contacting the vendor directly. Blackboard offers price quotes based on the number of students. Searching online for pricing is helpful for many VLE solutions, but Blackboard is among the vendors that don't release this information to the public.

5.4.2 Other Proprietary VLE Platforms

- *Canvas* – the developer and publisher of Canvas is *Instructure, Inc.* which is an educational technology company based in Salt Lake City, Utah. There is an option suitable for higher education. *Instructure* doesn't release pricing for Canvas. However, the cost to implement Canvas varies based on size, training, support, and other factors.

  Canvas charges a one-time implementation fee and an annual subscription fee based on an institution's total number of users. Canvas offers a free account for teachers, as well as a two-week, free trial with full-feature functionality. The free teacher account doesn't contain all the options available to paid Canvas users. Canvas also offers an open-source option for those who want to install it on their own servers instead of using Canvas' cloud.

  Users report that to get pricing information for the Canvas VLE one must contact the company. A quote and further details are available only upon request [5].

- *Edmodo*, https://new.edmodo.com/
  This is a social learning system and is reported not to be suitable for higher education [6].

- *Joomla* – this platform is derived from the Australian Mambo VLE. www.joomlalms.com/buy-now.html#hosted.

  The cheapest license for a cloud-based service is $400 per year for 1–50 seats and the minimum on-premise (with own server) is $299 for 100 seats (this is the minimum seats one can buy). The prices are public for their additional services and products –www.joomlalms.com/buy_now/. The reviews on internet specify as typical customer types of Joomla: Small and Medium Businesses, Large Enterprises, Freelancers. No reviews are available to the author's knowledge by higher education users.

- *Brightspace*, developed by **D2L** – (www.d2l.com/). D2L is a global cloud software company with offices in the United States, Canada, Singapore, Australia, Europe, and Brazil. The Brightspace VLE system is a cloud-based software used by schools, higher education, and businesses for online and blended classroom learning. As with previously mentioned VLE systems, Brightspace does not publish its pricing on its website, but internet discussions [7] provide the following information (2019): "services from similar providers have an average monthly cost of $452.25. Those with basic functionalities can range from $0 to $300/month, while the extensive ones can cost $259 to $1,250/month." A couple of users of D2L also report in these internet discussions that Brightspace has a rather steep learning curve.

It is noted that similar to other proprietary providers, prospective customers need to contact Brightspace (www.d2l.com/en-eu/contact/) for a pricing quote.

5.5 THE CONCEPT OF OPEN SOURCE AND SECURITY

Finally – security is one issue that makes people reluctant to use open source software. There is a lively discussion on the subject on the internet and in specialised literature, which goes outside the scope of this book. In the author's experience, working for over 15 years with WebCT and Moodle, and for 25+ years in e-Learning, security of both proprietary and open source VLE platforms is good and of comparable standard, and no significant issues have been encountered with either. Most users report good security for Moodle, if the code is not modified too much and if self-enrolment of students is avoided. Both of these requirements can easily be followed by the steps described in this book. However, for educators considering to implement Moodle, it is very important to adhere to all advice Moodle is giving on the issue of security – see https://docs.moodle.org/37/en/Security.

BIBLIOGRAPHY

1. Tsapaki, V., Tabakov, S., and Rehani, M. Medical physics workforce: a global perspective. *European Journal of Medical Physics* 2018; 55: 33–39.
2. Tabakov, S. Global number of medical physicists and its growth 1965–2015. *Journal Medical Physics International* 2016; 4 (2): 79–81.
3. www.betterbuys.com/lms/moodle-pricing/.
4. www.immagic.com/eLibrary/ARCHIVES/GENERAL/IMM/I060108F.pdf.
5. www.betterbuys.com/lms/canvas-pricing/.
6. www.betterbuys.com/lms/reviews/edmodo/.
7. https://fitsmallbusiness.com/brightspace-user-reviews-pricing/.

CHAPTER 6

Conclusion

THE AIM OF THIS BOOK is to provide a condensed guide on using Moodle by a professional without specific IT skills. The author has based the illustrations and explanations on real and working examples and has used her educational knowledge of Medical Physics and Engineering MSc delivery as the background of these examples. Highlighting the general principles in creating these examples makes the book useful also to a broader range of teaching scientists in other areas.

The experience with e-Learning over 20 years has convinced the author that it provides a quick, effective, and efficient way to transfer knowledge. However, this experience has always blended e-Learning with classical teaching. The presentation of parts of the author's work in the field of e-Learning at various conferences organised by the Consortia of the pioneering e-Learning projects described in the book confirmed that this view is shared by colleagues worldwide.

Since 1999, Medical Physicists and Engineers have been at the forefront of applying, developing, and spreading e-Learning. This was one of the major factors underpinning the double growth of the profession in the past 20 years, compared with the previous decades. Many colleagues have developed specific e-Learning materials, simulators, and interactive examples – now used worldwide.

Considering the management of MSc-level programmes with e-Learning, the author's experience spreads over 15 years and covers the use of platforms both proprietary and free and open source. In this experience, Moodle proved to be a reliable and most suitable VLE platform for our professional needs. It is also a platform with a smooth learning curve and the author believes that the short, condensed guide in this book will be useful for many colleagues who would like to start with or continue using Moodle. This will underpin one of the main challenges before Medical Physics and Engineering in the coming years – quickly educating young specialists to support the needs of healthcare.

Although the book is orientated towards Medical Physics and Engineering, we believe that it will be useful to many other specialists taking their first steps in Moodle.

Since the book was submitted for publication in December 2019, a global challenge has appeared, which placed life under quarantine, and required education institutions to seriously reconsider the way that knowledge is imparted. Their reaction is admirable and has humbled the author by its scope and quick response. To mention just two examples which the author has witnessed: the introduction of Moodle-based distance learning as a sole tool for instruction for all students at King's College London well before any formal directive; and the opportunity to ensure distance learning education of 6 to 18 year-olds in an entire country through the purpose-built platform shkolo.bg. The reader will of course be aware of numerous other examples where virtual learning platforms have been implemented at important times.

It becomes clear now more than ever that better understanding of e-Learning will ensure the success of future education models.

Index